A classic creation of British humour

T0174460

THE RETIREMENT OF A.J. WENTWORTH

THE WENTWORTH PAPERS, BOOK 2

H.F. Ellis

This edition published in 2019 by Farrago,
an imprint of Prelude Books Ltd
13 Carrington Road, Richmond, TW10 5AA, United Kingdom

www.farragobooks.com

By arrangement with the Beneficiaries of the
Literary Estate of H.F. Ellis

First published by Geoffrey Bles in 1962

Ebook ISBN: 9781788421812
Print ISBN: 9781788421843

Cover illustration by Ian Baker

Have you read them all?

Treat yourself to the complete Wentworth Papers –

The Papers of A.J. Wentworth, B.A.
In which he explains the truth of his misfortunes as a schoolmaster at Burgrove.

The Retirement of A.J. Wentworth
Officially retired to his rural village home, surely he can now stay out of trouble?

The Swan Song of A.J. Wentworth
Marking the end of an era, he dons his cap and gown one last time.

And turn to the end of this book for the chance to receive **further background material** on the series.

Introduction

It is now a good many years since my friend A. J. Wentworth, B.A., labouring under a sense of grievance, allowed me to publish some account of his experiences as a schoolmaster at Burgrove. His intention at that time was to defend himself against ill-natured gossip arising from an incident in which he threw an Algebra book, not of great weight, at one of his boys.

The attempt was not altogether successful. In the intervening years, Mr Wentworth tells me, he has received a number of letters – some abusive, some contemptuous, some even from Americans – which clearly show that his character and outlook on life ('my philosophy of living' is his own phrase) have been widely misunderstood. In particular he resents the suggestion, made by several correspondents, that he is just a 'typical narrow-minded schoolmaster', with no ideas or interests beyond time-tables, square brackets and Matron.

Now in retirement, 'A. J.', as he allows me to call him, feels it his duty, if only in fairness to his profession, to counter these damaging misconceptions. 'I make no explicit defence or apologia,' he finely writes. 'It appears to me that a simple, straightforward account of my life, as I live it in retirement from day to day, should suffice to show that, for variety and breadth of interests, civic sense, tolerance and a readiness to meet and mingle with all sorts and conditions of men and (up to a point, of course) women, a retired schoolmaster can hold a candle to any Tom, Dick or Harry, be he clergyman, stockbroker, peer of the realm or even a so-called "executive" like Sidney Megrim – *Megrim*, eh? – whose dog continues to foul the footpath outside my house despite my utmost endeavours.'

So he has sent me a great mass of personal jottings, extracts from diaries, letters, photographs, cuttings from local papers, theatre-ticket stubs, travel literature and (for some reason) a bulb catalogue with 'Narrow-minded, my foot!' scrawled across it. 'Do what you like with these,' he instructs me, 'as long as you give a fair picture of my life and philosophy of living, both here at Fenport and on my, I think I may claim, fairly extensive travels.' All these documents arrived wrapped up in a copy of the *Fenport and West Acre Advertiser* containing a red-ringed report of Mr Wentworth's acquittal on a charge of driving a motor-mower without due care and in a manner whereby a breach of the peace might have been occasioned.

I have done my best to carry out Mr Wentworth's instructions. But I am not entirely sure that he has been well advised.

At Fenport

A young fellow called at my cottage this morning to ask whether I would care to become a Vice-President of the Fenport Football Club. Monday is the day on which Mrs Bretton takes my laundry home with her and, of course, I was expecting nothing of the kind.

'Come in, come in,' I said. 'There's nothing else that I know of, Mrs Bretton, thank you. I put it all together in the bathroom.'

My visitor said he was Ernie Craddock from the garage and happened to be passing. I am old enough to remember the days when he would just have stood there, twisting his cap about awkwardly in his hands – not Craddock of course, but a much older man of the same age – and saying nothing. But times have changed. 'Wash day, eh?' he said. 'Shan't keep you a tick, though. The thing being, what about you as a Vice-Pre of the club?'

I said, when I understood what it was that he was trying to say, that it was a great honour to be asked and one that I greatly appreciated. 'What would my duties be exactly?' I asked, and he told me that I should not have to do anything actually. 'It's just the subscription, see,' he explained.

'I see,' I said. 'Well, I shall have to think it over – Yes, Mrs Bretton? What is it now?'

She is an excellent creature in many ways, and cooks my lunch and so on except on Sundays, when her husband is at home, but she does not understand that there are times when one does *not* wish to be bothered with household matters.

'No, not those,' I said, to be rid of her. 'The thin blue stripe this week, please. I was never a great footballer, I fear, Mr Craddock,' I went on, 'so that if there were any question of taking an active –'

'No refereeing, eh? Is that it?' he cried, breaking out into loud laughter, in which I confess I did not join. I suppose he intended a reference to a trivial incident that occurred towards the end of the cricket season, when I was unexpectedly asked to umpire a local match. The whole thing has been ridiculously exaggerated. It is nonsense to say that I deliberately tripped up one of the Fenport batsmen in the middle of the pitch and then gave him out. I have umpired hundreds of games at Burgrove Preparatory School in my time, including one or two First Eleven fixtures when Rawlinson was away, and such a thing has never happened before. I had a perfect right to cross over from one side of the pitch to the other, as soon as I noticed that a left-hander had arrived at my end – indeed it was my duty so to do.

'The bowler was entirely to blame, Mr Craddock,' I said with some warmth, when he had had his laugh out. 'He had no business to deliver the ball without first ascertaining that I, as umpire, was in a position to adjudicate if called upon to do so – as in fact I was.'

'The bowler, eh?' he repeated, breaking out afresh. People often laugh a great deal, I have noticed, when they realize they are getting the worst of an argument.

'Certainly,' I said. I had no doubt in my own mind that the bowler was to blame. The result of his precipitancy was that in crossing the pitch I inadvertently collided with the batsman, who was running a leg-bye, and to add to the confusion he, in falling, made an involuntary sweeping movement with his bat and brought down the batsman from the other end. Meanwhile the stumps had been thrown down and, on appeal from the wicket-keeper, I had no option but to give the decision 'run out'. Even, as I explained later to the Fenport captain, had I been in any way to blame, there is no provision in the Laws for 'Obstruction of the striker', whether by an umpire or anyone else. It was simply one of those instances of bad luck, which sometimes occur in cricket and must be accepted in a sporting spirit.

'I thought I should have died,' Craddock said, beating his hands against his knees in a rather affected way. 'The three of you lying there, and you on your back with your finger up calling "Out!". "OUT!" says you. "Out! Who's out?" says somebody. "Both of 'em?" And then, on top o' the lot –'

I enjoy a good joke as well as any man, but I fail to see the humour in what was, after all, a straightforward cricketing dilemma. It is true that, on being asked amid some laughter which of the two batsmen I had given out, I declined to give a ruling; but my difficulty was that both of them – indeed, all three of us – lay at a point roughly equidistant from the two wickets, and I was unable without reference to *Wisden* to say how the Law stood. I therefore, rightly I think, referred the matter to my colleague at the bowler's end, who called very belatedly, 'No ball!' and ordered the game to proceed – a highly improper decision, in my opinion. Be that as it may, no one had a right to criticize my own handling of the incident who has not himself been faced with a similar situation on the cricket field – at short notice too, remember.

I was endeavouring to point this out to Craddock, and growing a little heated at his continued silly laughter when Mrs Bretton, perhaps fortunately, came down again to say that she had found my spectacles in 'the upstairs place' as she prefers to call it.

'I have my reading glasses here, thank you, Mrs Bretton,' I said, taking them out of my pocket.

'These'll be the others then – for looking out the window and that,' she replied. 'Though what you'd be wanting with them in the upstairs –'

'Never mind that now,' I said sharply. 'Just leave them on the table, please.' I have no wish to hurt her feelings, but I really do not feel that I can enter into long discussions about where and why I have left this, that or the other thing. I had enough of that at Burgrove, towards the end.

Craddock took himself off soon afterwards, remarking that a guinea was the usual thing and adding that he had had 'a wonderful time' – a glib phrase, meaningless in the circumstances, which I suppose he has picked up from his betters or the wireless or somewhere. I saw him to the gate and was just too late to shout at Sidney Megrim's infernal terrier. He pays no attention whatever to the by-laws relating to footpaths, and I shall really have to speak to him about it. To Megrim, that is to say. It is useless to speak to the dog.

When I had dealt with that, it was time to go down to the village for tobacco, and with one thing and another my *Telegraph* crossword had to be left till after lunch. Not that it matters, in a way. I am no stickler for a rigid routine, I am thankful to say. It is just that one likes to have the afternoon free for other things.

A letter from Gilbert, who will be following me into retirement in another five years or so, I suppose, set me thinking about the old days after tea. Or rather, set me thinking after tea about the old days. One likes to be accurate – I have my mathematical training to thank for that – without being too fussy, and as a matter of fact it was after breakfast I happened to be thinking about. This being Monday the Stationery Cupboard should have been open between 9 and 9.30 a.m., and I could not help worrying. Gilbert has had a lot of experience, of course; he knows about the pink blotting-paper for Common Room use only, naturally. But experience isn't everything, now that he has taken over full responsibility. One has to be pretty firm at times, with a lot of boys clamouring round for pens and rubbers and so on, to which very often they have no right. For want of a nail, they say, the battle was lost, and the same thing applies, up to a point, to nibs and Common Entrance. Of course, it is none of my business now, and nobody I dare say is indispensable, difficult though it may be at first for my colleagues to realize it. 'We struggle along as best we can,' Gilbert very kindly writes, 'without you.'

It is a relief, in a way, to be free of the worries and anxieties of a senior master at a preparatory school, but life without responsibility and duties, as I told them all at the Presentation, is an empty thing. ('As empty as a swimming bath without any water in it,' I could not resist adding, a reference which they greatly appreciated to the time when, through an absurd misunderstanding, I opened the waste-pipe on the morning of the School Diving Competition. A joke against oneself helps to relieve the tension, I have often found, even on a serious occasion; though my heart was heavy enough, naturally, as I stood there with my right hand resting on the mahogany bureau subscribed for by many old Burgrovians, as well as past and present colleagues.) 'I hope,' I said to them, when the laughter had died down, 'that I shall find no less useful work to do at my new home in Hampshire.' This, for some reason, started the younger boys laughing again – through over-excitement, I think.

I certainly intend to take as full a part in the life of the village as my reading and other tasks and interests allow. But these things take time. One cannot become a churchwarden overnight. I should have thought that any fool would understand that. I thought so at least until this evening, when I sauntered down to the post office to get my reply off to Gilbert.

I was standing quietly at the stamp counter, waiting until the sub-postmistress had finished selling a pair of canvas shoes, when quite by chance I happened to overhear a conversation between two men at the far end of the shop, one of whom I recognized by his voice to be Mellish from the chemist's.

'Who's that little old chap went up street just now, then?' the other man asked.

'Struts along?' Mellish asked. 'Shortish, and looks over the top of his specs? Name of Wentworth. Elm Cottage. Didn't you hear?'

The other man said no, and Mellish then gave a very inaccurate and one-sided account of the umpiring incident to which I have already briefly referred. I had half a mind to go across at once and point out the fact (which Mellish had deliberately omitted) that it was the bowler who had been to blame. It would have served Mellish right to be made to look foolish in public. But, to tell the truth, I am sick and tired of the silly business, and besides that I strongly object to eavesdropping in any shape or form. I should have made my way out of the shop, had there been any way of doing so unobserved. As it was, I was forced to remain and listen to the rest of the conversation, distasteful though it was.

'What is 'e, then, when he's not playing the giddy-goat?' the second man asked. (Giddy-goat, indeed! My old friends at Burgrove would have something to say about *that*.) 'What's 'e do?'

'Do?' Mellish said. 'Nothing.'

'How's he manage that, then?'

'Easy,' Mellish said. 'Starts right in at it after breakfast, so I'm told, and keeps on without a break till bedtime. A proper gentleman, old Mother Bretton calls him.'

'I should like a threepenny stamp at once, please, Mrs Enticott,' I said loudly. 'I am in rather a hurry, as I have a great deal to do.'

I propose to leave the matter there. At my age one has learnt to treat idle, ill-informed gossip with the contempt it deserves. All the same, I shall be surprised if Mellish does not take a very different view of my character before the year is out.

A Misunderstanding
at the Greengrocer's

I am not at all sure that time is really saved by this modern habit of abbreviation. The longest way round is often the shortest way home, as my old nurse never tired of telling me. She is dead now, of course, but the principle of the thing remains. 'Just you put one leg in at a time, Master Arthur,' she used to say. 'Then we shan't have you toppling over backwards into all that Plasticine.' She had a fund of wisdom, rest her soul, and the world would be a better place if there were more of her like about today.

Her words came back to me at the greengrocer's this morning. I mean about the shortest way home, naturally – not the other. At my age one doesn't topple over backwards at the greengrocer's or anywhere else; or at least, if one does (and the truth is, whatever Miss Edge might say, that the place was abominably crowded and I for one would rather step a little too close to the carrot rack than inconvenience a lady any day), the reasons are different. What made me think of my old nurse was this silly trick of saying 'three' instead of 'three pence' or 'three-pennyworth'. Surely we are not all in such a rush to get wherever we are going that we have time only for monosyllables? The result in any case is to waste time as often as not, as happened this morning when I said I wanted some brussels sprouts and asked the price.

'Ten,' the man said.

I naturally supposed he was inquiring how many I wanted and told him it was just for myself, to go with a chop, and thought perhaps eight would do, if they were large ones. When I am more used to shopping for myself I shall know in a flash, but at present I am rather feeling my way.

He is one of those sandy-haired young men, always trying to do six things at once. 'Not up there, Fred,' he shouted. 'They're under the caulies. What did you want, then?'

'About eight, I think,' I began, but broke off because he was telling a lady that the lettuces she was fingering were fresh in today. Then he took fourpence from a thrustful woman who came up with a lemon, and began to shovel up vast quantities of sprouts in a scoop.

'Got a sack?' he asked me.

'Good heavens!' I said. 'Are those for me? All I wanted –'

'Make up your mind,' he said. 'Eight pound, you *said*. If it's for Mrs Odding,' he added, speaking over his shoulder to a girl in a white overall, 'she has the Cos.'

I began to lose patience with these constant unmannerly interpolations. 'I do not care tuppence, young man,' I said warmly, 'whether Mrs Odding has the Cos – or the flu either, for that matter. All I want is a little civility, and some sprouts.'

One of those odd silences fell over the shop while I was speaking, and it may be that a sudden consciousness that heads were turning in my direction put me momentarily off my guard. At any rate, in making way for a woman who came bustling past me with a push-chair, I took an incautious pace backwards and fouled the carrot rack. It is absurd to have such an insecure structure in a busy shop. After all, one can move about perfectly freely in a fishmonger's without bringing a hundredweight of haddock about one's ears; or in a hat shop or chemist's for that matter, *mutatis mutandis*, now that they no longer pile up sponges and loofahs in inadequate wire baskets. There ought to be more consideration and common sense. Ironmongers hang their surplus shovels and brooms and so on from the ceiling, and though that might not do in every case it shows what can be achieved by the exercise of a little imagination. At Burgrove Preparatory School the boys' bowler hats (for travelling, etc.) used to be kept piled up on a shelf at elbow level, but soon after my arrival there as Assistant Mathematical Master they were moved to a high cupboard out of reach – another case in point.

Everybody was very kind and helpful, but in the end I left without buying any sprouts and went, almost directly, to Gooch's for tobacco. He also sells walking-sticks, though I don't quite see the connexion. Something to do with the open air perhaps, unless briar was once used for both – I mean for pipes as well as for walking-sticks, or rather the other way round. But that hardly seems likely.

Miss Edge was in Gooch's and shook her finger at me in a way I do not much admire.

'What's this I hear about your pelting Mrs Odding with carrots, Mr Wentworth?' she said immediately.

The speed with which gossip, and highly inaccurate gossip at that, flies about a small place like Fenport still astonishes me, though I have been there six months or more now and ought to be finding my feet. It was bad enough in a Common Room, but here!

'Pelting Mrs Odding! 'I cried, hardly able to believe my ears. 'Why, I –'

'With carrots,' she repeated, nodding. 'Over at Wright-son's. *And* calling her names, by all accounts. I didn't even know you knew her.'

'I do *not*,' I replied, colouring up. 'What is more, Miss Edge, sorry as I am to scotch so succulent a snake at birth, she was not even in the shop at the time. I merely – '

'Oh, Mr Wentworth!' she said. 'Behind her back! That *was* naughty.'

'One does not pelt people with carrots behind their backs,' I began heatedly; but noticing that Mrs French and her little boy had entered the shop and seemed to be listening I broke off and asked for an ounce of Richmond Curly Cut.

'Five,' the girl said.

'No, one,' I corrected her.

'Five shillings,' said Old Gooch, intervening. 'Just gone up.'

Five shillings! And I can remember when it was seven-pence. Still, there it is. One must move with the times or go under, as happened to a poor old friend of mine when haircuts went up to one-and-six. After all, it is cheaper to have one's teeth out now than it was in the old days, so that one thing balances another up to a point.

Miss Edge was nowhere to be seen when I turned to continue our conversation, and I made my way home in a thoughtful frame of mind. It is idle to concern oneself overmuch with the small contretemps of every day; gossip-mongers will make mountains out of molehills, do what one will. None the less it distressed me to think that this Mrs Odding might be led to believe that I had spoken rudely about her, if (as was more than probable) some garbled version of the incident were to reach her ears. It would be better, I decided – rightly I still think – to ring the lady up and explain the whole thing quite simply, before Miss Edge or anybody else had a chance to upset her with ill-natured tittle-tattle. But of course that meant that I must waste no time at all.

There is only one Odding in the book – it is an unusual name, I think – so there was no difficulty about that.

'Odding here,' said a man's voice, when I got through.

'I have just come from the greengrocer's,' I explained,' – that is, I wonder, could I speak to Mrs Odding, please?'

'For you,' I heard him say. 'Chap from the greengrocer's. Got a mouthful of potatoes to get rid of, by the sound of it.'

If a lifetime's schoolmastering has taught me nothing else (as it certainly has), I have at least learned to disregard trivial rudenesses. Of course, in this instance I was no doubt not intended to overhear what was said, but I could not help wondering whether Mr Odding has ever paused to consider what his *own* voice may sound like when distorted by the telephone.

' Yais? What then? Mrs Odding spiks,' another voice said.

'Oh, Mrs Odding,' I said – with rather a sinking heart, to tell the truth – 'I am sorry to trouble you, but I just wanted to clear up a small matter, a silly little incident at the green –'

'Is arrived,' Mrs Odding said. 'He is come O.K. Beets and all.'

'Yes, yes,' I said. 'This is another matter. I happened to be in Wrightson's this afternoon when one of the assistants –'

'Is right?'

'Wrightson's, Mrs Odding. With a W, you know. One of the assistants, the sandy-haired one actually, mentioned that you preferred the Cos, the lettuce, you understand –'

'Alwais,' she said firmly. 'Never the other. It is becoss of the wind.'

'I see. Well –'

'Up she comes, else. With Cos, no. If he is not Cos, back she goes. You know me?'

'No, Mrs Odding,' I said. 'That is what I wanted to explain.' I did not, as a matter of fact, want to do anything of the kind. I am not in any sense a Little Englander: some of my best friends are Balts and Slovenes and so on; and I am well aware of the importance of reaching a close understanding with people who have not had the same advantages as we have – or perhaps one ought to say not the same kind of advantages, to avoid the risk of misunderstanding. But, really! When it comes to clearing up

20

a silly little affair in a greengrocer's one would rather have to do with one's own kidney.

'Mrs Odding,' I went on, speaking slowly and distinctly, 'I simply rang up to ask you to take no notice of any stupid stories you may hear about a trifling incident in the greengrocer's this afternoon. When I tell you that people are already going about saying that I pelted you with carrots, Mrs Odding, at a time when, as you and I know perfectly well –'

'Here, I've had about enough of this,' said Mr Odding's voice. 'Who the devil are you? And what do you mean by trying to frighten my wife with a lot of damn balderdash about carrots? Ringing up in the middle of tea and scaring a woman out of her wits after all she's been through these last months –'

I allow no one to take that tone with me, least of all when I am attempting to make an apology.

'My name is Wentworth,' I said coldly, 'and I would have you know that –'

'Aha!' he said. 'So *that's* it.' And rang off.

Two minutes later the phone rang again, and I supposed it would be this man, Odding, come to his senses and anxious to explain himself. But it turned out to be Harcutt, a solicitor whom I have met once or twice at the library and so on.

'I say, Wentworth,' he said, 'speaking as a friend, is it true you called Miss Edge a succulent snake at the chemist's this afternoon?'

One really has no patience with this kind of folly.

'If you are thinking of joining the Old Women's Scandal-mongering Society, Harcutt,' I suggested, 'you had better try to get *some* of your facts right. In the first place, Miss Edge and I met not at the chemist's but at Gooch's.'

'That certainly alters the situation,' he said.

'And secondly, I called her no such thing. I should have thought you knew me well enough by this time. The amount of petty gossip and trouble-making that goes on in this place –'

'Well, keep your hair on, Wentworth,' he said – an expression I have always disliked. 'I was only joking.'

'And while we are about it,' I told him, 'here is a further bit of information for you. I did not throw carrots at Mrs Odding in the greengrocer's either.'

'You didn't?'

'No.'

'Then why mention it?'

'Because I have no doubt that it is all over the village by now.'

'Wait a minute, Wentworth,' he said. 'Are you suggesting that things have reached such a pitch that the fact that you *didn't* throw carrots at somebody is red-hot news?'

'Oh, go and boil yourself, Harcutt,' I said. He is a good chap in many ways, but I was not in the mood for that kind of schoolboy facetiousness. All this fuss over buying a few sprouts, and even then I did not get any.

I took up the dictionary after supper, to calm my mind, and looked up 'briar'. Apparently they don't make pipes from the prickly kind, but from the root of a sort of heath, which makes the connexion between tobacco and walking-sticks all the more mysterious. The word comes from the French *bruyère*, to my surprise. I had always thought it a kind of cheese.

Brains Trust

The Conservative Association holds from time to time what they call a Brains Trust. St Mark's Hall is not a very cheerful place for it, in many ways; the walls are dark green and tend to sweat, and of course rows and rows of rush-bottomed chairs never suggest cosiness. Still, people come. I dare say they like to see our local M.P., Sir Arnold Bantry, who is always on the 'panel', and other notabilities from Fenport and West Acre and even farther afield. It makes an evening, as Mrs Wheeler puts it.

I little thought, when I first strolled along to St Mark's six months or so ago, that I should one day be 'on the panel' myself. But Mrs Dalrymple has been most pressing. 'Everybody's tired of these endless barristers and journalists,' she told me. 'We need fresh blood, a new outlook. Do say yes.' At first I demurred. One does not like to stand in the way of the younger people. Besides, my knowledge of the world is to some extent departmentalized and, though I make it my business to take a lively interest in everything that goes on around me, I might well be floored by a question on, let us say, the Middle East oil situation or – as I told Mrs Dalrymple – anything about marital relationships, which I have never had. However, she entirely misunderstood my point.

'There's nothing to be frightened of, Mr Wentworth,' she said. 'The audience are very friendly and the Chairman will help you all he can – Sidney Megrim is *so* good – and see you safely through those "beginner's nerves".'

The idea that I, after a lifetime's schoolmastering, might be alarmed at the thought of facing a roomful of middle-aged ladies in a tuppeny-ha'penny place like Fenport was really too much. I made a somewhat emphatic reply, which Mrs Dalrymple took as an acceptance, so that I found myself committed before I had really considered the matter fully. 'Next Thursday, then, at 8 p.m.,' she said gaily. 'And the very best of luck.'

The short notice made me wonder whether somebody else had fallen out at the last moment, and the suspicion became a virtual certainty when the Chairman introduced the members of the panel to the audience, getting my name right but adding that I was an authority on Elizabethan Drama and had been in my day a well-known amateur actor. No doubt somebody had forgotten to alter the notes given to him about the original panellist. It is all in the day's work, but even so! 'In my day', if you please. One is not a hundred and fifty. I have never acted, as it happens, but if I had I dare say I could act as well today as ever I did. Better, very likely, with all the added experience of life and character and so on that the years bring. Everybody clapped politely, as they do at these affairs, and the Chairman read out the first question before I had a chance to put the matter right.

It was about woman's proper place being in the home and Sir Arnold, who spoke first, made the apt point that it was the Conservative policy to encourage women to devote to their homes the time and energy demanded by their proud positions as wives and mothers, while at the same time developing in the larger world outside those public and civic qualities upon which so much depended. He then criticized the Socialists for their attitude to the Early Closing Act of (if I remember rightly) 1922, and was reminded of an amusing incident in which he missed an important appointment in London through some misunderstanding about spaghetti. I forget how he linked it up, but it was all very skilfully done and showed how a practised speaker can turn the most unlikely subject into grist for his mill.

A Mr Philip Tallboys said he agreed with every word the Member had said, and had little to add except it was no good saying that woman's proper place was in the home unless you first made sure that she *had* a home to take her proper place in. The Conservatives had no reason to be ashamed of their record in providing houses for the people. He is a chartered accountant, with a habit of slouching down in his chair and tapping his finger-tips together that struck me as a shade patronizing at a public meeting of this kind. Miss Gorman, on the other hand, who followed him with an interesting account of her life in a biological laboratory, leaned right forwards with her hands clasping her bag and described the 'worthwhileness' (to use her own phrase) of dissecting water-beetles and so on with an eagerness to which the front row of Fenport housewives were slow to respond. Or so I thought. It is a capital mistake to try to win sympathy by over-emphasis. Let the facts speak for themselves – such at least was my method when teaching geometry at Burgrove – instead of repeatedly urging their importance and interest.

While Miss Gorman was speaking I caught the eye of a man about four rows back who was putting a sweet of some kind in his mouth. Old habits die hard and I gave him a sharp look. I suppose there is no definite rule about eating at public meetings, but it would look very odd if those on the platform started putting toffees in their mouths, and what is sauce for the goose ought surely, as a matter of ordinary politeness, to be sauce for the gander. At any rate the man I caught at it clearly felt uneasy, for he gave a little cough and put his hand up to his mouth again – an old trick, designed to make one think that that was what he was doing the first time. Thereafter, whenever I glanced in his direction he instantly stopped chewing and stared straight ahead with rigid jaws. It was annoying to be unable to tell the silly fellow that a man with my training could see through a dozen better dodges than that. I remember a boy called Mason who took the rubber out of the metal holder on the end of his pencil and substituted pieces of mint humbug, cut to fit. He is an aeroplane designer now, they tell me, and putting his ingenuity, one hopes, to more worthwhile ends. But he very nearly fooled me at the time.

I mention this small incident only to explain why my attention was momentarily distracted when the Chairman called my name.

'Eh? What's that?' I asked.

'We are hoping,' he said amid some laughter, 'that you would have something to say on this question.'

'What question is that?'

'We have been discussing,' the Chairman said, with a long-suffering air that rather nettled me, 'whether woman's proper place is in the home.'

'Have you, indeed,' I replied. 'I was under the impression that you had been discussing whether the proper place for water-beetles was the Conservative Party.' This sally, which was of course intended, in part at least, to be in jocular vein, was very well received by the audience. Sir Arnold, however, saw fit to take offence.

'We certainly get all sorts at our meetings,' he remarked.

'It is quite clear, at any rate,' put in Miss Gorman, 'that the proper place for a woman is not Mr *Wentworth's* home.'

'I entirely agree,' I said. 'As I happen to be a bachelor, it would be most im proper.'

'That is a state of affairs that could easily be remedied,' the Chairman pointed out, when the laughter had died down. 'Perhaps Miss Gorman would co-operate?'

'Oh, I'm afraid my time is very much taken up with my water-beetles, which Mr Wentworth seems so much to despise,' Miss Gorman said, with a rather tight-lipped smile. 'Somebody else must accept the honour.'

'No room for just one more?' Tallboys asked.

I enjoy the cut and thrust of debate as much as any man, but there is a point, a very definite point, at which good-humoured raillery ends and impertinence begins. Miss Coombes, who looked after the boys' vests and so on at Burgrove, put it very well, I remember, when she said 'Anyone can be clever, Etheridge, but it takes a gentleman to be courteous.' Though I said nothing, some of the annoyance I felt may very well have shown in my face. At any rate, the man in the fourth row, in whose direction I happened to be looking, hastily whipped his handkerchief up to his mouth and unless I am very much mistaken ejected whatever he was eating and put it into his pocket. Thereafter, I was glad to see he paid close attention to what was being said, and once or twice nodded his head to show agreement with some observation of my own. I know the type from of old.

A man like Tallboys, with a hide three inches thick, has to be handled differently. It is best, as a rule, to ignore them; and that is what I did. The Chairman in any case had gone on to the next question, which concerned our relations with the United States. Miss Gorman said that the future of the world depended on a close accord between the English-speaking peoples, and with this we were all agreed. Then we discussed juvenile delinquency, a subject on which I suppose I can claim to speak with some authority. Not, naturally, that there was anything serious of that kind at Burgrove, but boys are very much alike the world over and only those who have spent a lifetime in their company can hope to have a close understanding of their problems. I said that provided young people were brought up in the right way, taught to fear God and be honourable and straightforward, tell the truth and play the game in the widest sense of the phrase, we need have no fear that they would grow up in the right way. The audience took this very well, I think, and Sir Arnold backed me up by saying that a good home was the important thing. Miss Gorman talked about glands and suchlike hocus-pocus, and Tallboys said he did not think a boy was any the worse for stealing a few apples when the policeman wasn't looking. If that was how he spent his own boyhood, he is not, in my opinion, much of an advertisement for apple-stealing, but of course I did not say so. I leave cheap scores to others. The Chairman summed up, neatly enough, by saying that the panel seemed to be agreed on the whole that the younger generation would be all right provided they were taught to behave themselves.

After that we discussed interplanetary travel and mixed marriages and the best present to give a man on his fiftieth birthday. Then the Chairman announced that there were five more minutes to go and asked if any member of the audience had a question he or she would like to put. At once a woman got up and asked the extraordinary question 'Can the panel explain why there is no Request Stop at the bottom of Penfield Road, and isn't it high time something was done about it?'

Well!

I suppose it is natural, in a way, that people's small local concerns should be of more interest to them than some of the questions, many of them of fundamental importance, we had been discussing. But the immediate quickening of interest took me, I must confess, by surprise.

'Quite right, Mrs Burfitt,' somebody called out. 'Four hundred yards if it's an inch my little girl has to walk –'

'It's the shopping,' a thin woman in a beret explained. 'You don't want to trudge right up Elm Street with a heavy basket.'

'I hardly think –' the Chairman said.

'Take Friday. There was a good half-dozen of us in all that rain, and it isn't as if you could always take your overshoes. You'd think they could do *something*.'

'*Didn't* it come down, Mrs Enticott! I saw you there, but not to speak to, being in the back, and I said to my husband "The poor thing *will* get wet, I told him – "'

'It's the same with the footpath back of the common,' a bald-headed man complained, without troubling to rise to his feet. 'You could break a leg in one of those pot-holes. It's all very well for the panel to come here and talk about mixed marriages and satellites and that, but what I'd like to know is when is something going to be done about the disgraceful state of some of our public paths?'

'Hear! Hear!' said several voices.

What astonished me was the Chairman's acquiescence in this absurd situation. He seemed perfectly content to sit back and allow the meeting to degenerate into a disorderly hubbub. Of course in a way it was no business of mine, but when a young woman got up and demanded two additional hard tennis courts in the Recreation Ground, or some such nonsense, I really could stand it no longer.

'This is not a meeting of the Rural District Council, madam,' I said. 'The panel is here, if the Chairman will allow me to say so, to answer questions of general interest, not to repair footpaths or insist on additional Request Stops –'

'But a stop at the bottom of Penfield Road *is* of general interest,' interrupted Miss Gorman, in what I am afraid was a rather obvious attempt to curry favour. 'It is certainly a matter of great interest to *me*.'

'In that case,' I replied, unable to repress a touch of sarcasm – a weapon that I am normally very loth to use, 'I shall of course make it my business to take the matter up with the bus company at the earliest possible opportunity.'

'And don't forget the footpath,' somebody called out.

These people seem to think that, just because one is on a platform, one has nothing better to do than see to all the trivial problems they are too lazy to work out for themselves. Really, one might as well be back at school.

One Thing after Another

Word has got round that I am taking up the matter of the Penfield Road bus stop. Mrs Wheeler went out of her way to congratulate me about it this morning after church.

'We need somebody of energy and initiative in this place, Mr Wentworth,' she told me. 'There is so much to be done. It is splendid to know that you are going to interest yourself in our small concerns.'

It is absurd that a chance remark, made in a spirit of irony at a so-called 'Brains Trust', should be taken up in this literal way. But what is one to do? Mrs Wheeler would rightly have felt very rudely rebuffed had I replied that I had no intention of doing anything of the kind. Besides, I suppose in a way it is just the opportunity I have been looking for to give a helping hand here and there in Fenport. When one joins a community one has no right, even in retirement, to sit back with folded hands and expect others to do the work. 'Heaven helps those that helps theirselves' is an old adage of my nurse's that comes to mind.

I was revolving in my mind the opening phrases of a letter to the West Acre and District Transport Company, when I was hailed by Miss Stephens from the Bank. She is petite and does not, I am sure, dye her hair. Some women have a natural bronze that grows a little paler at the roots.

'Oh, Mr Wentworth,' she began in her breathless way, 'of course we hardly know each other, and I do hope you will forgive me, but the fact of the matter is it's the Dramatic Society. We are doing *The Linden Tree*, you see, it's a Priestley of course, and the old Professor – he's out of date, if you remember, and *won't* give up despite his wife's longing to get away from it all –'

'But, Miss Stephens,' I interrupted gently, 'I am rather at a loss. I shall of course be delighted to take a ticket when the time comes –'

'But we want you to *act*, Mr Wentworth,' she cried. 'It's a part that's simply made for you. So gentle, and yet so firm. I can just see you, in the big scene with Mrs Linden –'

I naturally supposed at first that she was pulling my leg.

But after a while as she prattled on, I realized that she seriously imagined I might be persuaded to go on the stage in front of half the people in Fenport and involve myself in this scene with Mrs Linden, whoever she might be. At my age! So I seized the first opportunity to tell her that I had never acted in my life, and that I feared I was a little old to begin now.

She opened her eyes very wide indeed.

'But, Mr Wentworth!' she said. 'I know that you used . . . before half the crowned heads of Europe, Mr Megrim told me.'

'Oh, Megrim!' I said.

I saw at once that this, like the bus stop business, was another result of Sidney Megrim's stupidly inept handling of the Brains Trust meeting. The next thing would be, no doubt, that somebody would ask me to give a talk to the Literary and Debating Club on Elizabethan Drama.

'Let me assure you, Miss Stephens,' I said, 'that any stories about my having once been an actor are quite unfounded. Apart from one occasion in the old days when I dressed up as Father Christmas for an end-of-term concert, I can truthfully say –'

'And what are two of my parishioners hatching up together against me now?' cried the vicar, coming up unexpectedly from behind. He has a way of taking one by the elbow that, in one not of his cloth, I should be inclined to resent. 'Something to my detriment, I'll be bound.'

Somers is a good man, with no high-church nonsense about him, but perhaps rather roguish for his years.

'You are barking up the wrong tree, Vicar,' I told him, seeking unsuccessfully to free myself. 'I was simply explaining to Miss Stephens that, though I have dressed up as Father Christmas in my time

'The very man,' he interrupted. 'The very man! Now that poor old Witherby has been laid to rest. Miss Stephens, you shall persuade him. Four o'clock in the Church Hall, on December the 20th. Vestments provided by the parish. Splendid, splendid. You have no idea, Wentworth, how difficult it is nowadays to get the right sort of person to take an interest in our small doings.'

'Well, really –' I began.

'Mr Wentworth is thinking of joining our Dramatic Society,' Miss Stephens said. 'Just for a small part, perhaps, until we see how he gets on. We mustn't ask *too* much of him, must we?'

'Mine is no more than a walking-on-part, as we say,' said the vicar. 'Just a word here and there to the kiddies. A pat on the head, perhaps. You have patted heads in your time, eh, Wentworth? Four o'clock then, on the 20th of next month.'

'I very much fear,' I said – but he had darted off in his impetuous way to accost a young couple with a dachshund, and I turned to find Miss Stephens laughing gaily at me.

'*What* a breezy Christian it is,' she exclaimed. 'I always feel I ought to hold my skirts down when he's about.'

I dug the ferrule of my umbrella into the ground, at a loss for a reply.

'And hold my hat on tight and so on, I mean,' she went on with a becoming blush. 'That *was* naughty of you, about him barking up the wrong tree. I very nearly burst right out.'

'Naughty?' I said. 'In what way?'

'Don't pretend you don't know everybody calls him Frisky Fido. I should think he knows it himself, very likely.'

'Oh, that,' I said. 'Oh well.' It was the first I had heard of it, as a matter of fact, but I cannot say I was greatly surprised. The more I see of life in this neighbourhood, the more it seems to me to resemble life in a boys' preparatory school. The gossip and childishness of it all, and now nicknames for the vicar! The only difference in a way is that there is practically no discipline.

We had quite a pleasant talk, until our ways parted at the china shop. Miss Stephens is an amusing little thing, and I should be glad to help her, within reason. But there are limits. I do not think I have committed myself about her absurd play-acting suggestion, though of course, up to a point, it might be a new experience. We shall see.

It is odd to remember that a little time ago I was complaining of a lack of useful employment here in Fenport. And now here I am, practically rushed off my feet with this Penfield Road business and rehearsals for Miss Stephens's Dramatic Society starting on Wednesday (it is really impossible to refuse so pressing a lady), not to mention the vicar's Children's Party, though that of course is not for a few weeks yet. Then there is the Vice-Presidency of the Football Club – I mean I have still to settle whether to accept the honour at a guinea a year or send them a gracefully-worded refusal. All these things mount up.

'Don't you go and do too much, Mr Wentworth,' Mrs Bretton said to me only this morning as I was on my way out to the tool-shed to straighten things up in there. 'You're too kind-hearted by far. You want to take it easy, your time of life, not run this way and that for a pack of women.' She is an excellent creature, but a little inclined to treat me as though I were an elderly invalid. 'That Miss Stephens,' she added, and would I think have said more, had I not shown by my manner that I had work to do. I do not believe in encouraging gossip, however well intentioned.

I keep my bicycle in the tool-shed, together with the mowing machine and other odds and ends. This is inconvenient at times, because of the way things have of catching in things when you want them – or rather when you don't. I mean when you don't want the things they catch in, naturally, and it is these for some tiresome reason that generally come out first when you pull. Besides, it is bad for the spokes. A lot of old paint-tins and so on were left behind by the last occupant into the bargain, and the place needs a thorough tidy. I hate anything slipshod. The School Museum was in my charge at Burgrove for a number of years, so I am not without experience of arranging a number of diverse objects to the best advantage in a narrow space.

The right way to begin is to clear everything out first, and this I proceeded to do (not, for reasons that I have explained, without some difficulty). I shall never understand why my predecessor here needed so much garden hose. It is awkward stuff at the best of times, and really, for the small lawn and two rose beds which are all I have here, one would have thought forty feet or so would be enough. I very much doubt, as a matter of fact, whether he ever used it at all; parts of the middle section have clearly, to judge by the cobwebs alone, been entangled in an old wooden rake for many years. At any rate, after a tug or two, I decided that the only thing to do was to unthread the hose carefully from end to end, to disengage it, that is to say, from the various articles with which it had become entangled. These latter could then be removed *seriatim*.

To my astonishment, the hose appeared to have no end, whichever way I traced it, and it was while trying to follow a double loop through the framework of a deck-chair that I somehow got the handle of a pair of edging shears up the left leg of my trousers. I should not ordinarily note down the details of so mundane an operation as tidying out a shed, but there has been a lot of exaggerated gossip in the village – *nothing* is too trivial, it seems, for some people – and injustice to myself I wish to explain, quite briefly, what happened next.

Anyone who has tried to trace an unrolled garden hose, or a tangled fishing cast for that matter, to its source knows that it is essential at no time to lose touch. The eye is not to be depended upon. The entire length must be followed *by hand*, for otherwise it is all too easy to reverse one's direction unknowingly where two loops intersect and thus arrive back at one's starting point. For this reason I was obliged to force my way *through* rather than to go *round* the deck-chair, and by a stroke of ill-luck the serrated, or notched, leg of the chair, which happened to be uppermost, swung over on its pivots and trapped the upper part of my body. At the same time the point of the shears became jammed rather awkwardly in the lower framework – of the chair, that is – and in attempting to preserve my balance I inadvertently gave a sharpish tug to the hose (of which, I must repeat, I could not let go without throwing all my trouble to the winds) and brought down a quantity of sacking and old dahlia tubers from a shelf above my head. I was temporarily blinded. But for that I should certainly not have made what turned out to be a false move.

I suppose it may have been two or three minutes later that I heard footsteps on the gravel path outside. Not wishing to be interrupted I kept quite still, and it was with some irritation that I heard Mrs Bretton call out, rather gruffly, 'Here's Miss Stephens wants to see you, Mr Wentworth.' Still, there was nothing to be done, and a moment later I could sense that Miss Stephens was standing in the doorway.

'I'm so sorry to bother you when you are busy,' she began. Then her voice died away. ' Oh my God, Mrs Bretton!' she cried. (I suppose it is the modern way, but I do *not* like it in a woman.) 'There's something hanging up in the corner.'

The truth was that in an instinctive effort to free myself by slipping backwards through the framework of the chair and at the same time *shrugging* it, if I make myself clear, upwards, I had caught the leg-rung, by a million-to-one chance, on some nail or projection in the wall. Movement of any kind was now painful, if not positively dangerous, and it was while quietly thinking out the next step that I had become aware that I had a visitor.

'It is perfectly all right, Miss Stephens,' I said, to reassure her. 'I was looking for the end of the hose.'

'Goodness, you gave me a shock,' she said. ' It's hanging down your back.'

'Aha!' I said. 'It must have been hidden among the dahlias.'

'So are you, you poor thing,' Miss Stephens said, coming – very expertly, I must say – to my assistance.

The upshot of all this is that I have practically promised to take a part in her play. Apparently, a Wally Bishop is going abroad, and they are desperate for somebody to act the part of Lockhart, whoever *he* may be. To tell the truth, I was not in the best of tempers – nobody likes to be interrupted in the middle of a worthwhile task – and the quickest way to settle the matter seemed to be to say yes.

'In that case,' I said, 'I suppose – very well. By all means, if you wish it. Yes.'

'You *darling!*' Miss Stephens said. Of course, I know that it means nothing, among stage people, but all the same one was glad that Mrs Bretton had gone to get her husband's lunch.

The first performance is to be on December 20th, apparently. I have a feeling that I already have some engagement for that date, but Miss Stephen says it doesn't matter. I don't know, I'm sure, whether I have been altogether wise.

The Penfield Road Affair

A man named Willis, who is something to do with the Gas Company I believe, called about the H-bomb this morning. He told me that sheep were eating grass coated with Strontium 90, or some such number, and wanted me to protest about it.

'This is all a little bit outside my province,' I told him. I know nothing about sheep, and was in any case rather busy.

'Is the extermination of the whole human race outside your province, then?' he asked.

'It is certainly more than I have time for this morning,' I said, ignoring the rudeness of his manner. 'At the moment I have my hands full with this business of a Request Stop at the bottom of Penfield Road. Perhaps, while you are here, you would care to join in a protest against the intransigent attitude of the bus company? Only a few days ago, I am told –'

He seemed to be a man without any sense of proportion. 'Penfield Road!' he cried. 'A fat lot it will matter whether the buses stop at Penfield Road or anywhere else when there's nobody left to ride in them and the whole of Europe is a desolate waste. It's people like you, with their noses buried in their own petty little local affairs, that are bringing the world to the brink of destruction. Can't you realize that already seaweed is being dredged up with point nought six of a fatal dose of radiation in it? And you talk about bus stops!'

I am not the man to be hectored on my own doorstep.

'Listen to me, young man,' I said. 'When I was your age there were plenty of people going about saying that the end of the world was at hand. They used to hold meetings, I remember, with the slogan "Millions Now Living Will Never Die". A fine state we should all have been in if I had listened to them and decided it was not worth while to go on teaching my boys trigonometry. Get on with your own job, my lad, and leave sheep and seaweed to wiser heads.' For two pins I would have told him to get his hair cut and take his bicycle-clips off before calling uninvited at strange houses. I did, in fact, say something of the kind, more or less *sotto voce*, and he went away muttering.

I had meant to get a strong letter off to the bus company (or the *Advertiser* perhaps would be better), but this Willis interview unsettled me and after thinking things over for an hour or two I set off to change a book at the library. Mrs Wheeler was there, half-way up a ladder, and we chatted for a while.

'I feel like Romeo and Juliet,' she said. 'Why not come and have dinner one evening, Mr Wentworth? On Friday week? One or two people will be there whom you might like to meet, if you haven't met them already.'

'I'm sure I shall be delighted to meet them even if I *have* met them already,' I said politely. I had not intended any joke, but Mrs Wheeler began to laugh so I joined in.

'That would depend, wouldn't it?' she said. 'Suppose I asked Mr Willis?'

'Willis!' I said. 'You mean the gas man!'

'He says you threw his bicycle-clips in his face. I must say it doesn't sound very likely, but he *did* seem angry. Of course, he did not say it to me, but they were outside Gooch's just now and I couldn't very well help hearing. Then that man Odding –'

'Odding?'

'Yes. They were talking, you see. Odding said nothing would surprise him. He said you rang his wife up the other day and threatened to pelt her with carrots. There was something else, but I had my shopping to do, and of course –'

'This is getting beyond a joke,' I broke in angrily. 'If those two men are getting together to spread slanderous stories about me I shall certainly take action. The whole thing is a mare's nest. I simply rang up Mrs Odding to explain that if she heard any silly stories about my – about carrots being thrown at her at the greengrocer's it was all a misunderstanding. She is unfortunately an Esthonian –'

'But, Mr Wentworth, surely even an Esthonian would know whether she was or was not being pelted with carrots. I mean if it wasn't true, I don't quite see –'

'Exactly,' I said. 'The carrots have been trumped up against me by ill-disposed gossips. There were a few on the floor, not more than a few, and nobody was hurt. Actually, the whole thing started with a lettuce.'

'You threw a lettuce at Mrs Odding?'

'I neither threw, nor threatened to throw, anything at anybody, Mrs Wheeler. I merely gave vent to an expression at the greengrocer's –'

'Muriel!' Mrs Wheeler called out suddenly. 'Mr Wentworth is telling me how he gave vent to an expression at the greengrocer's.'

I had not noticed Miss Stephens come into the library. She now joined us with an 'Oh, do tell!' and the two ladies listened sympathetically while I gave a short account of the circumstances that had led me to make a hasty remark in Wrightson's shop about Mrs Odding's order for lettuces, and how my attempt to apologize to the woman on the telephone had been misinterpreted by her fool of a husband. 'The whole thing is a storm in a tea-cup,' I ended.

'I don't suppose you really threw Willis's cycle-clips in his face either,' Mrs Wheeler commented.

'Oh *no*?' Miss Stephens cried. 'Mr Wentworth, you really *are*!'

I explained that the man had no doubt been speaking figuratively. 'I may be old-fashioned,' I said, 'but I would never dream of going to anybody's front door without removing my bicycle-clips, and I do not expect people to come to mine without removing theirs. All the same, I should have said nothing, of course, if the man had been civil. Standing there lecturing me about seaweed and decrying the Request Stop, when I am old enough to be his father, though I am bound to say that I should have to be a *great* deal older before I would dream – however, that is neither here nor there. The man was uncivil, and I sent him away with a flea in his ear, as my old nurse used to say.'

Both ladies were blowing their noses, and there was a short silence.

'I'm sure he richly deserved it,' Mrs Wheeler said. 'He was rude about the – about the Request Stop, you say?'

'I thought he would be better employed in joining me in a protest about the Penfield Road business than in badgering people with a lot of nonsense about radioactive sheep. "Get your hair cut," I told him, "and leave all that tomfoolery to wiser heads".'

'That settled *him*, I should think,' Mrs Wheeler said. 'But I *am* so glad you are taking the Request Stop so seriously. And what a splendid idea to organize a protest!'

'Hardly that. Hardly that, dear lady,' I said. 'I am merely writing a letter to the *Advertiser*, which I think will do some good.'

'I certainly mustn't miss *that*,' Miss Stephens said.

Their interest and enthusiasm encouraged me to go straight home and write a fairly stiff letter to the local paper, without further delay. It will make the West Acre and Fenport Transport Company sit up, I fancy.

This Penfield Road bus-stop affair is becoming more of a nuisance than it is worth. A day or two after posting my letter to the *Advertiser* it occurred to me to walk out and see the actual terrain, what was involved in the way of distances between existing stops, etc. 'Get out into the field, Wentworth,' my old C.O. used to say to me in the last war, and though I never quite became used to being spoken to as if I were a horse, I am sure his advice was sound. An ounce of knowledge is worth a deal of theory, as they say.

It was a considerable shock to me to find that there is already a Request Stop at Penfield Road. I can only suppose that the bus company had got wind of the fact that the matter was being taken up by someone who was not likely to be put off by excuses and evasions, and had decided to anticipate what they knew to be an unanswerable demand. I felt very badly let down, and my first thought was, of course, to withdraw the letter I had written to the *Advertiser*. This meant fourpence in a telephone kiosk which I can ill afford.

'I wrote you a letter the day before yesterday,' I said as soon as I was through. 'About the Request Stop at Penfield –'

'Do you want Basting?' a voice said.

'Certainly not,' I replied. 'I wish the letter to be withdrawn immediately. It must not appear.'

'You want the Editor then,' the voice said. 'Only he's out.'

'Who is that?' I asked sharply. 'I must speak to the Editor. This is a matter of the utmost –'

'Mr Basting's gone, see? If it's about the letter, it's in. This is Partridge 'ere, the boy, and it'll be out Friday.'

'No, no,' I said. 'It must come out *now*! The letter must not appear. I am speaking from Penfield Road, and the fact is that a bus stop has recently been installed. That being so –'

'We knew that, o' course,' the boy said.

'Then naturally you will not print my letter, which was written in ignorance of the facts.'

'It's in,' he repeated. 'On the machines. "It'll stir up correspondence, anyway," Mr Basting said. "There'll be plenty of people glad enough to point out the error," he said. He said we don't get a letter putting its foot in it right up to the neck *every* week, he said. "And what's more," he said –'

Never, even in my schoolmastering days (except perhaps in the matter of the changing-room pegs in Poole's time) have I met so utterly irresponsible an attitude. 'Listen to me, Partridge,' I said. 'I think you are mistaking your man. You may tell your Editor from me that if my letter is printed I shall not hesitate to write him another, which he will not like. He has no business to include correspondence, written as I say, in ignorance of the facts –'

'Look,' the boy had the impertinence to reply, 'if we cut out letters just because they were written in ignorance of the facts, we wouldn't *have* no correspondence. "Bung it in," Mr Basting said to me –'

I was not prepared, of course, to listen to this kind of talk. 'Bung it in,' indeed! Sometimes I wonder what the world is coming to, when young flippertigibbets, scarcely out of their teens to judge by their voices, can speak in that strain to a man old enough to have forgotten more than they ever learnt. If this is what comes of trying to give a helping hand to the people of this village they will soon find that they have a very different kettle of fish to deal with. It is all very well to pester me to dress up as Father Christmas next December and have this extraordinary scene with Mrs Linden, if that was the name, and do this that and the other thing, but if all I am to get in return is a lot of inaccurate gossip about throwing carrots, at *my* age, and now this trouble with the *Advertiser*, I might as well go back to Burgrove and try to hammer a bit of sense into a lot of boys who, with all their faults, never dreamed of using strong language in my presence. There is such a thing as flogging a dead horse, as I shall tell them.

The Party at the Vicarage

It has been a trying day, and might well have ended in a contretemps. But as I have often found in life, difficulties and mishaps are there to be overcome, and if faced with calmness and common sense sometimes turn out to be all for the best. Looking back on it all, I cannot feel that there is anything to regret, though of course another year, before agreeing to give away the presents at the Vicarage Christmas Party, I shall make certain that it does not clash with the opening night of the Dramatic Society's play – assuming (as I am bold enough to do) that Miss Stephens will again ask me to take a part. Dashing from one engagement to another, for all the world as if I were a Prime Minister or some such functionary, is a bit too much of a good thing at my time of life.

Things would have run more smoothly, I dare say, but for a muddle over my Father Christmas outfit. This was no fault of mine, for I was in good time at the Vicarage and naturally expected to find everything ready for me there. It was quite a shock when Mrs Somers greeted me, almost before I had got my bicycle-clips off, with a cry of 'But where are your *things?*' She told me that they had been sent round to my house so that I could try them on, get used to the feel of the beard and so on – 'We always did that in poor Witherby's time,' she said, as though that had anything to do with it – and what she could *not* understand was how I had failed to see the parcel.

'Naturally I saw the parcel, Mrs Somers,' I replied. 'At least, I saw *a* parcel that somebody must have left on my doorstep when I was out. But I am afraid I am old-fashioned enough, when parcels come for me at this season, not to open them until Christmas Day.'

Mrs Somers threw up her hands in a gesture that, considering the circumstances, I found somewhat irritating. 'The children will be bitterly disappointed,' she said. 'We had hoped that you would be ready to greet them when they arrived, and – Oh dear! Here comes Mrs Thompson already with her two. *What* are we to do?'

Well, I have faced worse crises many a time in the old days, without losing my head as Mrs Somers seemed to have done. I shall not forget in a hurry the time the under-Matron came back from a wedding in a very strained state, and I had to make an immediate decision whether to send the boys out on a run or order them into the gymnasium for extra P.T. It would have been better, as it turned out, to send them for a run; but the important thing, as I explained to the Headmaster, was to get them out of the way quickly, not to stand about debating the pros and cons. I could not possibly know that Miss Vincent would herself go straight to the gym to sleep it off. There was the occasion, too, when the chapel organ became waterlogged on Confirmation Sunday – but there is no need to go into all that now. My point is that an old schoolmaster is not likely to be thrown out of his stride by a temporary hitch at a children's party.

'Surely, Mrs Somers,' I said quietly, 'it will be more effective if I arrive *after* the children are assembled? As though by sleigh. It will take me no more than twenty minutes to ride home, collect the parcel and return.'

'They will see you arriving,' she said. 'They are as sharp as needles.'

'It will be dark,' I replied. 'What is more, to be on the safe side, I will put on the costume before approaching the house. I do not see, in that case that it will matter even if –'

'Oh, Mr Wentworth!' she cried. 'Without a sack, and wheeling a bicycle! The children would be bitterly –'

'Then what do you suggest, madam?' I asked, very nearly at the end of my patience.

She bit her lip. 'Perhaps, if you wouldn't mind,' she said – 'Here comes Mrs Whitney's little boy, in tears already – if you would come in by the side gate when you get back? Down the lane here, and up through the garden to the conservatory door. You would be out of sight from the front, you see, and we could give you the sack –'

'Give me the sack, eh?' I put in. 'There's gratitude for you!' But she was too fussed to appreciate the joke, and with an abrupt 'I must fly!' went off into the house.

I found when I got home that the parcel, mainly because of the top-boots, was too bulky to be carried on my bicycle, so there was nothing for it but to ride back to the Vicarage in full Father Christmas rig. The beard I could of course have strapped to the carrier or thrust into an inside pocket, but after some hesitation I put it on. It would be better, I thought, in case I were seen on the journey, not to be too readily recognizable.

I have had little or no experience of bicycling in fancy dress, and after one or two unsuccessful attempts to mount in the normal manner I realized that it would be necessary to pull the cloak up about my waist. I therefore gathered the skirts and had raised them with some difficulty as far as my knees when a voice called out 'Oh! Oh!', adding as it receded into the dusk, 'Time for another Request Stop agitation.' I could not see the face of the passer-by who made this senseless remark, but I have reason to believe that it was Willis, of the Gas Company, a man I have never liked. He has some bee in his bonnet about saving the world from annihilation, but would be better advised in my opinion to start by mending his manners. At any rate I made no reply and once I had learned the trick of turning the toes of my top boots outwards to avoid fouling the front mudguard was soon on my way. I dare say some of my Burgrove boys would have been surprised if they could have seen their old master pedalling along in a white beard, with a red cloak tucked up round his knees; but I have never been afraid to do something a little out of the ordinary, I think I may claim, particularly if it is to save kiddies from disappointment.

The journey, as a matter of fact, was uneventful (though I thought I heard some silly giggling as I turned into Dyson Road), and I reached the Vicarage without accident. The front of the house was ablaze with lights, but the lane itself was very dark, which made it a matter of some difficulty to find the side gate. Still, it was not many minutes before I was safely inside the garden and, leaving my bicycle by the hedge, began to make my way cautiously up a narrow gravel path.

It would have been a kindness had Mrs Somers thought of posting somebody with a light to guide me in, and I can only suppose that her anxiety not in any way to disappoint the children led her to neglect this small attention. The result was that I must have branched off the direct route to the house and, after fruitless knocking, entered a greenhouse, which I naturally took to be the conservatory. I suppose I must have taken half a dozen steps before the strong smell of damp earth, the heat of the place (excessive for December), and the complete absence of light combined to make me realize my mistake. I immediately turned round, though not apparently through a hundred and eighty degrees, and stepped into a cactus – an awkward enough customer even for an unbearded man. This mishap made me,

I am ashamed to say, momentarily lose my composure, and in wrenching myself free I incautiously stepped backwards off the duck-boards, lost my balance and toppled, rather than fell, into a trough or container of some soft flourlike substance, which the vicar told me later was horticultural soot.

Like most men who have led a busy active life I have once or twice before found myself at a loss in dark confined spaces, and experience has taught me the vital importance of standing absolutely still. Not indefinitely of course, but long enough to allow the mind to formulate a plan of campaign. On this occasion I adopted the same course, and very soon decided that I had only to regain the duck-boards and, keeping my left foot in contact with their edge, follow them to the door. Sure enough, some half-dozen paces brought me to the end, and extending my right hand I was greatly relieved to find it in contact with what must be the handle of the door. To this I gave a half-turn, and was at once aware of a fine but persistent jet or spray of water which seemed to come at me from every direction. The explanation, that I ought of course to have kept my *right* foot in contact with the duck-boards, did not immediately occur to me, and I fear that I did some little damage in my anxiety to be gone, before I understood that I was now at the wrong end of the greenhouse. Thinking back over the incident I am inclined to agree that my wisest step, after that first instinctive leap backwards, would have been to search again for the controlling handle or tap and turn the water off. But men do not always, in emergency, do the wisest thing. What I very much resent is the suggestion that I at any time panicked or lost my head. The simple fact that, when on my way back down the greenhouse, I kept the soot-trough constantly in mind and on this occasion disentangled myself from the cactus with proper deliberation should be enough to scotch *that* snake.

That I was vexed and put out by the whole business I do not deny. Indeed, when I regained the dry open air my intention was to return straight home, however bitter the disappointment might be to the kiddies, and it was pure chance that led my footsteps to the conservatory instead of to my bicycle. Having got so far, however, I thought it my duty to let some member of the household know that I was unfortunately unable to give away the presents as arranged. Accordingly I made my way through the back portions of the house, where I saw nobody except a woman working at a sink, who gave a shrill cry as soon as I accosted her and locked herself into a cupboard. I have no patience at all with female hysterics and simply passed on into the hall, which was brightly lit and gaily decorated with streamers and balloons. From a closed door leading, as I knew, into the dining-room came the sound of youthful voices and the clatter of tea-cups, but the hall itself was empty, and it was while pondering what best to do that I happened to catch sight of my reflection in a large and, to my mind, rather rococo looking-glass.

I at once decided that my best plan, after all, would be to go quietly home and telephone from there. One is prepared, of course, to look odd, even ridiculous, in the garb of Father Christmas, but I was certainly not prepared to present myself in company in the state to which I was now, through no fault of my own, reduced. The gown itself was not seriously torn, except about the hem, but my face was so mottled and streaked with soot and perspiration that I do not believe even my old colleagues would have recognized me. Rivulets of blackened water had run down the full length of my beard, which was sadly awry. My hands, naturally enough, were coated with soot and dried blood (a good mixture I remember thinking – so oddly does the mind associate in times of stress – for brussels sprouts). There was leaf-mould not only on my boots but even on my scarlet hood, and twined about my right sleeve was some kind of creeper with curious fleshy leaves.

It was in trying, with an understandable gust of annoyance, to shake or flick this last encumbrance from me that my right-hand came by ill-luck into contact with an old-fashioned gong immediately behind me. A low resonant boom rang through the house, and I had scarcely time to take one quick step towards the front door when the diningroom door was thrown open and Somers himself stood framed in the aperture.

'St Chrysostom!' he said.

My first concern was to assure him that the sounding of the gong had been purely accidental. One does not, naturally, summon people intentionally in such a way – least of all in their own homes.

'I was flicking off a creeper –' I began.

'Flicking off a creeper!' he repeated, in the voice of a man who attaches little or no meaning to what he is saying. 'My dear fel – Not you, Jackie. Get back, boy, and shut the door.'

He was too late, however. A little red-haired boy, who belongs I think to the butcher, peered round his legs and set up a cry of 'He's come! He's here! Father Christmas is come down the chimley!' At once there was a rush from within the room, and a cluster of boys and girls pushed and jostled their way into the doorway. I realized that it would never do, whatever my personal vexation, to forget the part I was supposed to be playing. Accordingly, I did my best to smile cheerily at the youngsters, and gave them a welcoming wave of one of my blackened and bleeding hands. 'Merry Christmas!' I said.

One or two of the smaller ones started to cry, but I very soon put an end to that.

'Stop that noise this instant!' I ordered, and there was an immediate silence. I have not been a schoolmaster for thirty-five years for nothing – little though it was that I got, in another sense, I am bound to say.

Mrs Somers had by this time worked her way to the front of the group, and I heard the vicar tell her, still in the same low bemused voice 'All he *says* is that he was flicking off a creeper.' It was a difficult situation for her, in some ways, as for me. But she is a well-bred woman, and her years of training as a hostess told her instinctively what to do.

'Perhaps, dear,' she said briskly, 'Father Christmas would like a wash.'

'Of course, of course,' the vicar said, coming forward while his lady began to shepherd her charges back into the dining-room.

'What on earth, Wentworth?' he went on, when the door was safely shut. 'My dear fellow, you are all wet, apart from the – Where have you *been*?'

'I have been in your greenhouse,' I said shortly.

'In my green –? To shelter from the rain? I had no idea –'

'Not to shelter,' I said, and there was something in my voice that warned him, I think, that I did not wish to pursue the matter at present.

'Come along, anyhow,' he cried. 'Come along and wash. This way, this way. Mind the gong.'

'I am neither blind nor incapable, thank you,' I said irritably. 'It was by pure mischance that I had my back turned when shaking or flicking –'

'Shaking or flicking?' he said. 'Dear me! I hardly know –' He has a tiresome habit of rushing into speech before one has fairly begun what one has to say. 'By the *way*,' he added, halting suddenly in front of a door, 'I suppose when you had – er finished in the greenhouse, you remembered to shut the door?'

'I doubt very much,' I replied, 'whether it makes any difference now whether the door is shut or open.' I did not, to tell the truth, very much care.

'My calceolarias!' he cried – or some such ridiculous expletive. 'In that case – in here, please. Forgive me!' And he was off, leaving me to tidy up as best I might in peace.

Well, it had been an unlucky business in many ways. But misfortunes, as I have said, can sometimes be turned to good account. The children were not at all frightened when I finally entered the room carrying a laden sack. Indeed they clustered about me, demanding to know what it was like coming down a chimney; which one I had chosen; where my reindeer were, and suchlike childish questions. 'You really made the tiny ones *believe* in Father Christmas again,' Mrs Somers told me later on. 'You see, they had got rather too used to poor old Witherby. I do think it was clever of you to think of the soot.'

'Yes, yes. Yes, indeed,' Somers said. 'Though I could wish – if you had thought of asking – however, there it is.'

I thought it best to leave it at that for the time being. Everybody was most kind, and even when I gave little Felicity Bennings a pat on the head and a quantity of soot, which must have somehow lodged in my sleeve, fell on her flaxen curls, the incident passed off with general good humour. There were games and crackers, and really, despite my damp gown, the years seemed to roll away. It was quite a disappointment when people began to leave. I was playing 'Puss in the Corner' – at my age! – when Mrs Somers came running from the telephone to say that I was wanted at once at the Parish Hall.

'It's Miss Stephens,' she said. 'She says the curtain goes up in ten minutes, and where on earth have you got to?'

'God bless my soul!' I cried. 'The play! I had forgotten all about it.'

The Play

I hardly know how the play happened to escape my memory, except, of course, that I was late at the Vicarage on account of the imbroglio in the greenhouse and one thing and another. Goodness knows it has been in my mind often enough the last few weeks, with rehearsals three days a week, and now one of my two arm-chairs borrowed for scenery. I am to be Alfred Lockhart, who comes in first and says 'Oh – I say, is this right?' not the Professor. Lockhart is described in the directions – the play we are doing is, of course, *The Linden Tree*, by a man called Priestley, about whom I know little though I remember once reading what seemed to me a very much too highly-coloured account of life in a preparatory school by a man of the same name – at any rate Lockhart is described as 'a precise, anxious, clerkly, middle-aged man', which made me chuckle when I read it. 'Hardly the part for me, Miss Stephens,' I pointed out; but she said she was sure I could do it very nicely. I suppose that is the point of acting, in a way. To be somebody different, I mean. If we all went on the stage to play ourselves it would be just like everyday life, which nobody wants to pay half a crown to see, I take it, or even a shilling farther back.

I dare say Sidney Megrim has had a lot of experience in producing, as we call it, but his manner is sometimes a little, well, abrupt, considering the difference in our ages. It is not after all as if one were being *paid* to make oneself out to be 'precise and anxious' and all the rest of it – as is the case, for instance, when a younger Headmaster (*mutatis mutandis*, naturally) takes advantage of his position to administer a rebuff. At the very first reading of the play I had no sooner said this opening line about 'Oh, I say, is this all right?' (We were not in costume then, of course – strictly speaking: though as a matter of fact I wear my ordinary suit when we are – but even so I held up both hands as I spoke, to show surprise and doubt) – I had no sooner spoken than Megrim interrupted with a quite gratuitous 'No, it is *not* all right, Wentworth. It's terrible.'

'Indeed?' I said. 'And in what precise manner –'

'Look,' he said. 'You have just been shown into the room by Mrs Cotton here –'

'By Miss Stephens,' I corrected.

'By Miss Stephens, if you prefer it, who is taking the part of Mrs Cotton, woman-of-all-work to the Linden household. Right? You have come to see Mrs Linden. You expected therefore to be shown into the drawing-room. You find yourself in Professor Linden's study. Right? What's the good of gabbling off your opening line and *then* looking surprised? Come in. Look round the room. Register surprise – "Oh!"-and off you go.'

'You mean exit?' I asked. 'At once?'

'Off you go with your speech, man. Now try it again, there's a good chap.'

I said nothing, but at once, as a good trouper should, did my best to follow his instructions by looking round the room – St Mark's Hall, of course, actually – with growing astonishment.

'Get on with it, man!' Megrim said. 'You haven't come to take an inventory of the furniture.'

It is almost impossible to get the feel of a part if one is to be constantly interrupted.

'I have been shown in a good many wrong rooms in my time, young man,' I began –

'I dare say,' he said. 'But not into this one. Again, please.'

I kept my temper, of course. As Miss Stephens said later, it is all part of the game. 'You did it splendidly, Mr Wentworth,' she told me. 'Sidney only makes us go through it over and over again because it helps to fix the lines in one's memory.'

'I see,' I said. 'Yes. I hadn't thought of that.'

When the telephone call came through to the Vicarage to say that the play was due to begin in a few minutes it was a considerable shock to me. I am a firm believer in punctuality, and of course it is most important that all the actors should be present on the opening night of a new play. New to Fenport, that is; the play has already, I understand, been performed elsewhere. But by a stroke of luck a Mrs Downing offered me a lift in her motor to the Parish Hall, and I climbed in, still in the Santa Claus outfit in which I had been helping to entertain the children. There would be time enough to change when I got to the Hall. At least, not time enough exactly, but one knows what one means. The important thing was to get there without delay, and I am afraid I cried out a little impatiently when one of Mrs Downing's youngsters went back into the house for a missing balloon. 'It was a *red* one' the boy kept saying, as if that mattered. Children have very little sense of proportion at times.

Megrim was standing at the back, or I suppose I should say 'stage' door, and addressed me at once, characteristically without bothering to say good evening.

'My God!' he said. 'Where have you *been*? We've had to ring up.'

'I know that,' I replied, overlooking the blasphemy. 'I was at the Vicarage when the call came. I am sorry, but –'

'The curtain, man, the curtain!' he cried. 'Muriel Stephens is on now, dusting round and gagging. They got impatient. Get on as quick as you can and never mind the make-up.'

'At once!' I cried, and ran past him through the dressing-room.

I cannot think how I came to forget that I was still wearing the absurd red gown and rather soot-stained beard in which I had passed the earlier part of the evening. Looking back on it, I suppose the rush, and my anxiety not to keep Miss Stephens and the audience waiting, momentarily disturbed my judgement. One must remember, too, that I was already 'in costume' and so not unnaturally felt ready to 'go on'. Even a more experienced actor, I dare say, might become confused if he had to rush at a moment's notice from one engagement to another. I do not acquit myself entirely from blame, but as a fair-minded man I consider it was a part of Megrim's duty, as producer, to say a word of warning about my clothing. Be that as it may, I lost no time in making my entry and, after adjusting my eyes to the glare of the footlights and taking a quick look round the stage, as instructed at rehearsal, repeated my opening line.

'Oh, I say,' I said, 'is this right?'

Of course, strictly, I should have been shown in by Miss Stephens, or 'Mrs Cotton' to give her her stage name, but as she was already there, dusting and so on, that could not be helped. In any case as it turned out, it did not matter. Miss Stephens turned round, duster in hand, and instead of replying with *her* opening speech (which runs, as a matter of fact, 'Right? It's as right as we can make it. Nothing's right now, nor ever will be, if you ask me,' and so can hardly be described as difficult to memorize), simply stood and gaped. No doubt she was surprised by my appearance, but as an actress of considerable experience – however, there it was. The audience had broken into loud laughter immediately on my opening line, so it may be that, even if she had taken her cue, she would not have been heard.

The situation was a difficult one for both of us. The laughter grew in volume and I felt completely bewildered. I am no stranger to merriment, for a schoolmaster's life is by no means the dull and colourless affair that some people make it out to be, nor am I without the means of quelling it when it threatens to grow beyond bounds. But on this occasion I felt quite at a loss – until happening to put my hand up to my chin, as is my habit when puzzled or upset, I encountered my beard and at once became aware how incongruous my costume must appear in a modern play, even at Christmas time. It was difficult to know what to do for the best. Somebody off-stage was shouting 'Come off the stage, you fool!' which did not help. But what particularly distressed me was to catch sight of Mrs Wheeler, a lady whom I respect and admire, sitting in the front row and looking over her handkerchief at me with eyes that, unless I am much mistaken, were filled with tears. To her, at least, I felt that some explanation was due.

'I was unfortunately delayed, Mrs Wheeler,' I began, stepping forward to the footlights. 'There was a slight imbroglio in the Vicarage greenhouse –'

I got no further, partly because I saw that Mrs Wheeler was doubled up as though in pain, and partly because at this point somebody, Megrim I suppose, pulled the curtains across and I had much ado to avoid becoming seriously entangled in the folds. So that was that.

We rang up again, as the saying goes, about ten minutes later, after Megrim had made a short speech to fill in the time while I was busy changing and making up. It all went off very well, I think. It was a little awkward, naturally, to have to repeat my opening line, with which the audience were by now familiar, but I carried it off by varying the intonation.

'Oh, I *say*!' I said. 'Is *this* right?'

'Well, it's better,' some fool at the back called out (I suspect Willis), and there was renewed laughter; but neither Miss Stephens nor I took any notice, and after waiting for the noise to subside, went quietly on with the play. I must say, everybody seemed to enjoy it. I make no claim to be an actor, but I confess there were tears in my own eyes during my final scene with Professor Linden (actually Major Thorpe, from West Acre), when we discuss the need for colour and vision in education. How true that is! Major Thorpe I thought excellent. So indeed were all the others, and it was a real surprise to me at the end to hear my own name called out first of all. The proper thing, of course, as the author was not present, would have been to call upon the Major, or possibly the producer, to take a bow. I took no notice therefore, affecting not to hear; but the shout was taken up in all parts of the hall, and the clamour eventually reached such proportions that I was reluctantly forced to step forward and raise a hand to stop it.

'Ladies and gentlemen,' I said, when I was able to make myself heard, 'I am at a loss to know why I am singled out in this way. This is my first appearance on any stage – or rather,' I added with a smile, 'in a sense my second, as a gentleman at the back has just reminded me. I hope it will not be my last.'

'So do we,' cried Miss Stephens: a remark which was greeted with such prolonged and good-humoured applause that I found it by no means easy, hardened as I am to such scenes by innumerable Prize-givings and so on, to restrain my emotion.

'It has been a great privilege,' I continued, 'to learn some of the elements of this great art in the company of so gifted and distinguished a – er – cast. To Miss Stephens, who first enlisted my aid in this enterprise; to our indefatigable producer, Mr Sidney Megrim; to Major Thorpe, for whose performance tonight I feel sure the knowledgeable and incisive dramatic critic of the *Advertiser* will find the right word, if right word there be; to Miss Edge, but for whose painstaking make-up –'

I had intended to conclude the list of those who had helped in one way or another with the production by adding 'to all these I owe a debt that I can never hope to repay,' or some such suitable, and indeed sincere, phrase. But the applause and clapping that followed each name as I mentioned it was so enthusiastic that I fear I lost the thread of my own remarks. Indeed, during the particularly prolonged (and I am sure well-deserved) burst of cheering occasioned by my reference to Miss Edge, our make-up artist, my mind unaccountably wandered to the earlier part of the evening and I was unable to restrain an involuntary exclamation of dismay.

'God bless my soul!' I cried. 'I left my bicycle at the Vicarage.'

There was so much laughter at this – in which, after a momentary bewilderment, I readily joined – that I thought it best to leave the matter where it rested and give way to Megrim, who took the opportunity to say a few words of thanks on his own behalf.

So ended a memorable evening. Everybody has been more than kind, especially to one who took, after all, only a comparatively minor part. Even Odding, a man whom I have perhaps judged too hastily, came up to tell me that I had given him the best evening of his life. I am too old a hand to take such exaggerated praise at more than its face value. Still, there is no denying that kindness, as one gets on in life, is very warming.

One has much to be thankful for. It has been a tiring day, not without moments of difficulty that might easily, but for a certain – 'knack', shall I call it? – that only years and experience can bestow, have ended in disaster. But the ending has been a happy one. I feel that I am beginning to take a full part, a rewarding part, in the life of the village that is now my home. One does not want to become sentimental, but I think I may claim that I am 'accepted' in Fenport, that I belong. And that, to an old man who has sometimes been lonely, means much.

My mind is at rest, what is more, about my bicycle. The vicar rang up to say that he had stumbled across it by his garden gate and had put it in the greenhouse for the night. So it is, in his own rather odd phrase 'more or less under cover'.

A Disappointing Start

It is not easy for a retired schoolmaster to live in the manner
to which he is accustomed (and *that* is no great shakes, in all
conscience) on a small pension and the few shillings saved from
a lifetime spent trying to knock the elementary principles of
mathematics into a succession of thick-headed Burgrove boys.
One does not expect, naturally, to be able to afford television
sets and all the other fal-lals that I am told are now necessities
– necessities, forsooth! – for millions of people who have
never heard of Pythagoras and could not solve a second-degree
equation to save their lives, but a man likes a pipe of tobacco
now and again, and really there are times when one hardly
knows where to turn with collar-attached shirts coming back
in a parlous condition from the laundry at one-and-fourpence
a time and baked beans, which I do *not* like, almost a luxury
now. Something ought to be done, though it is hard to say
what. I certainly have no intention of accepting charity, from
the Government or anyone else, at my time of life. Not that I
would have accepted it as a young man, I need hardly say.

All that, however, is beside the point. I am not in my dotage
yet by any manner of means, and prefer to plough my own
furrow so long as I have my health and strength. My hearing
is remarkably good, considering, and as to my eyesight not

much escapes me, as Mrs Bretton found to her cost only the other day when she forgot to dust behind the clock. At the same time I do not altogether care for the idea of taking some fiddling coaching or tutoring job down here in Fenport, where I have other interests and, perhaps I may say without immodesty, a certain standing. The answer for a man in my position seems to me to be some form of part-time occupation elsewhere, it might be as companion or adviser to some young man or family, just for two or three months in the year. I fancy, with my experience, I could be of some assistance in any one of a number of situations that come to mind – and I use the word 'situation' in its widest sense, of course. One has not held the post of Headmaster's right-hand man (which I think I may say I was in my later years at Burgrove, despite young Rawlinson's tendency at times to – well, push himself forward a little more than his qualifications or attainments warranted) without acquiring the administrative ability and, shall we say, *savoir faire* that could be of inestimable value in, for instance, arranging a tour abroad, buying tickets and so on, or in taking charge of whatever it might be while others were temporarily absent. That kind of thing. Difficulties constantly arise, as I well know, where there is need for someone absolutely trustworthy to take over until matters sort themselves out. At any rate, we shall see.

> Retired schoolmaster, B.A. (Oxon), wd consider short-term employment, up to 3 mths if interesting work. Posns of trust, organization, etc. Willing to travel, within limits. Accustomed sole charge individuals or gps. No agencies or divorce work.

I should not myself have thought it necessary to make the final provision, particularly in an advertisement for insertion in

The Times, but an old friend whom I consulted on this matter tells me that it is advisable, if one wishes to avoid unpleasant entanglements. I am not exactly certain what 'No agencies' means, as a matter of fact, but I can well believe that it is something I should not care to be mixed up in. One hears of detective agencies, for instance. I can put my finger on the guilty party as promptly as any man when it is a matter of paper-dart-throwing or that bizarre business of Matron's overshoes, and I dare say the knowledge of human nature one acquires as a schoolmaster would stand me in good stead if I were ever called upon to investigate more serious crimes. I had no doubt whatever in my own mind about the theft of Miss Stevens's ducks down here the other week. Foxes, indeed! 'That was a two-legged fox, if ever I saw one,' I told her, and would very soon have named the culprit had she not already been paid compensation by the Hunt secretary. Still, that was all in the way of friendship. It is a very different matter to stand about in shrubberies at so much the hour and jot down the numbers of young men who arrive in Jaguars. The numbers of their cars, that is to say. I shall certainly decline any offers of work of that kind.

One will have to feel one's way. Who knows? Perhaps by this time next week I shall be far away from my little cottage in Fenport, engaged in some employment of a confidential nature. The years seem to roll away. One is on the threshold of, if not adventure at least a change. That is the great thing. I might almost be a young man again, eagerly awaiting the start of my first term as an assistant master. We shall see, as I have already said.

There have been no replies as yet to my advertisement, apart from a suggestion that I might be interested in investing £200 in a second-hand furniture shop shortly about to open in South

Wales. I am not. I have not put aside fifty pounds a year from my earnings ever since I was twenty-five in order to provide Welshmen with dressers and mahogany chests-of-drawers. Other considerations apart, I know nothing about second-hand furniture shops except that I have yet to see a customer enter one. I remember, years ago, asking a colleague why it was that such shops always kept half their furniture outside on the pavement, but he did not know. He said it was the same with people who sold ladders and enamelled baths. 'But not butchers or chemists/ I objected, and he agreed. One would have thought there would be by-laws.

My copy of *The Burgrovian* arrived by the same post as this extraordinary suggestion and I was skimming through 'Notes and News' (young Phillips has been playing hockey for Sandhurst, I see, though they have got his second initial wrong again) when an odd thought made me smile. 'Mr A. J. Wentworth's many friends, past and present, will be interested to hear that he has gone into the second-hand furniture business' – what a scoop that would be for next term's issue! But it is not my duty, I am thankful to say, to provide sensational items for my old school magazine. The whole thing is not worth a moment's thought.

Megrim came in about the Debating Society, and I asked him (quite casually, of course, and without giving anything away) if he happened to know anything about second-hand furniture in Wales.

'You ask the most extraordinary questions, Wentworth,' he said. 'I imagine it is like second-hand furniture anywhere else. Huge wall mirrors and purplish bureaux with dangling brass handles. Why?'

'Oh, nothing,' I said. 'It doesn't matter.' I then tried to change the subject, but Megrim is one of those people who worry at a casual conversation like a dog with a bone.

'Suppose I were to ask you whether you knew anything about ironmongery in Northumberland,' he said. 'Wouldn't that seem to you a bit odd? Or no, as a matter of fact I suppose it wouldn't. *Do* you know anything about ironmongery in Northumberland, Wentworth?'

I told him, as far as I remember, that I knew nothing about ironmongery, or about Northumberland either for that matter. I said, quite politely, that I was not in the least interested in ironmongery and that, if everything was now settled about the Debate, I had some rather important letters to write. But you might as well give a hint to a rhinoceros.

'*I'm* not interested in second-hand furniture,' he said, with a look round my room which I very much resented, 'unless it's good, that is. But I *am* interested, naturally, in why you should be interested. In South Wales of all places. Of course, if it's a secret –'

'There is nothing secret about it,' I interrupted, well knowing what would happen in a place like this if anybody thought there was. 'It simply happens – a friend of mine was asking me about it. I am only sorry I bothered you with the thing.' I do not care for prevarication, but with a certain type of persistent bore one sometimes has no alternative.

'I see,' he said. 'I tell you what though. You know Roberts, up the lane?'

'Well?' I said.

'Well, I've got an idea he, or it may have been his sister, used to be in Antiques or something more or less in those parts. He might be able to help. I'll send the old boy along to you, shall I?'

'It is very kind of you,' I began, 'but –'

'No trouble,' he said in that off-hand way of his. 'I'm going that way in any case.' And he took himself off before I had a chance to tell him for goodness' sake to mind his own business. If only I knew of an ironmonger from Westmorland or some

65

such place to plague Megrim with I could soon put a stop to this nonsense.

Roberts came to the back door after tea to tell me he had something in the handcart outside I might care to look at. I can't say I was greatly surprised, for these people always seem to get hold of the wrong end of the stick. Rather than involve myself in a long explanation to the effect that I was not interested in buying furniture but in the second-hand furniture business (which I certainly am *not*), I accompanied Roberts to the gate in no very good temper.

'That's a Swansea piece, that is,' he said. 'As you'll likely know.'

It was a small chest of drawers, hardly fit even for a boy's dormitory, and quite useless to me. However, I offered the man half a crown to be rid of him. It did not seem fair that he should have had his journey for nothing, under a misapprehension that was probably not altogether his fault. But he was by no means grateful. He said it was a genuine antique. 'Why, it's worth more than that as firewood,' he told me.

'I dare say,' I said. 'But there's the labour of chopping it up.'

In the end, feeling a little ashamed of having allowed ill temper to rule my tongue, I gave him ten shillings for it, which I can ill afford, and asked him to put the chest in the tool-shed. It seems a poor return, so far, for my advertisement in *The Times*.

Perhaps tomorrow's post will bring me something a little more promising.

A Missed Opportunity

It is all very well to say that all experience is an arch, as Tennyson somewhere or other claims, wherethro' gleams that untravelled world and so on and so forth. I used often to recite the passage, I remember, in my younger days when things went a little awry – to myself, that is: one does not declaim poetry aloud up and down the corridors of a first-class preparatory school! – and was much comforted by it at times. We live and learn, I suppose, would be another way of saying the same thing. But I very much doubt whether Alfred Tennyson, with all respect to his memory, ever had such a day as I have had, with nothing to show for it but a torn trouser-leg and a pocket full of unwanted oats. All experience indeed! A hundred and fifty miles all told, and an umbrella riddled with small shot through no fault of my own, at the end of it. What kind of untravelled world gleams through *that* particular arch, one is tempted to ask.

Whatever it may be, I feel disinclined this evening, with what I fear may be a heavy cold coming on, to explore it. One begins very seriously to doubt whether the insertion of my advertisement in *The Times*, asking for short-term work in positions of trust, etc., was altogether wise. Quite apart from being pestered with second-hand furniture down here in Fenport, some of the offers of employment made to me through the post have been ridiculous. One does not become a Bachelor of Arts, I should hope, in order to exercise dogs from 2.30 to 4.0 p.m. every afternoon except Thursdays. Nor am I the man, as those who know me best will agree, to recommend hosiery to total strangers on a commission basis! There are times when I really think the world has gone mad.

Still, to be fair, as I always try to be, the post in search of which I set off to Wiltshire early this morning appeared to be very much more in my line of country and might indeed have suited me well, but for a chain of ill-luck such as I have rarely experienced. An opening as companion-secretary to a gentleman temporarily incapacitated by an accident while hunting is the kind of thing I am looking for, at twelve guineas a week with board and keep and travelling expenses refunded after the interview if unsuccessful. I had little doubt in my mind, as I walked to the station to catch the 8.45 local to Southampton, that provided this Colonel Ripley proved to be of a congenial cast of mind we should very soon come to terms. Nor have I any reason to doubt *now* that we should have done so, had I been permitted to meet the gentleman.

Little is to be gained by jotting down the details of this vexatious business. The milk has been spilt, in every sense of the phrase, and there is an end of it. But I owe it to myself to point out that had adequate transport been available at Stenshall nothing of the kind would have arisen. I was thunderstruck when a rather dull-witted porter there told me that no buses ran past the Manor House on Wednesdays. 'There was a car come to meet the 11.48 down from Bland-ford,' the man said. 'But I doubt it could have been for you, seeing you just got off the 12.6 up from Temple Combe. Not that Mrs Ripley didn't go back empty in a bit of a taking, at that.'

There seemed no point in explaining to the fellow that I had been badly advised by a ticket-inspector at Bournemouth West, so I simply asked him what I had better do. After some thought he told me in his slow country way that Grimley's van would be coming 'up street' in a minute or two and might be able to drop me 'there or thereabouts', as he was pleased to put it. Having no alternative I agreed to this curious approach to my future employer, and very soon found myself jolting along in a bucket seat beside a civil young man, who seemed genuinely sorry that he had not the time to take me round past the Manor gates. He was going 'sort of more along the back of their place, like', he told me (How strangely these people talk!) but would drop me at the nearest point, where I could cut across a couple of fields and up through the farm. 'Ten minutes,' he said, 'at the outside,' and late though I was it seemed the best thing to do. One cannot pick and choose, really, when nothing else offers.

I could not foresee, of course, that a heavy downpour of rain would catch me out in the open as soon as I had entered the second of the two fields indicated to me. Naturally I had my umbrella with me, but I most certainly did not wish to appear for my interview with sodden trouser legs and I therefore turned left-handed and made all possible haste along the hedgerow to a barn or shed which stood in the near corner of the field, intending to shelter there through the worst of it. It was here that I had my first stroke of misfortune. The building was very dark inside, and though I could just make out that it contained some kind of machinery I had no warning that there was need to exercise more than my usual caution until I happened, while shaking my umbrella, to engage the crook of it with what must I suppose have been a lever or handle. I gave no more than a slight tug, to free it, and at once noticed a whirring and clanking noise suggesting that some sort of mechanical operation had been set in motion. Then a considerable quantity of oats, a very considerable quantity, was precipitated over my head and shoulders from above.

This incident, though momentarily startling and confusing in the indifferent light, was not in itself, to one who is accustomed to life's ups and downs, more than a passing inconvenience, and I should not have mentioned the mishap had not its immediate consequences proved so vexatious. Oats are less easy to brush off the clothing than some other kinds of cereal, and though I did my best in the few minutes that elapsed before the rain stopped I suppose it was inevitable that I should leave what amounted to a trail behind me on resuming my walk. Be that as it may, I soon became aware that I was being followed by several farmyard fowls, which appeared from nowhere after the manner of these creatures. Their number and speed increased rapidly, to my dismay, so that by the time I had reached the farther hedge, beyond which the roofs and chimneys of what must be Manor Farm were clearly visible, they had become a serious embarrassment. I am not easily put out, but no one cares to arrive for an appointment attended by a flock of gaping poultry. I therefore made an attempt to drive them off with gestures of my umbrella, which I still think was the only sensible course in the circumstances. There was an outburst of cackling and one or two fowls rose into the air with the usual exaggerated loss of feathers, and I was preparing to take advantage of the diversion to slip through a gateway when I heard the sound of running feet and a man's voice from beyond the hedgerow shouting, 'Get over left there, quick!'

Not knowing to whom he was speaking I checked the swing I was taking at a particularly pertinacious hen (a White Wyandotte, I think) and unluckily lost my grasp on my umbrella, which flew into the hedge some distance to my right, where it lodged quivering. At once there was a further cry of 'There he goes!' followed by the roar of firearms, and I found myself temporarily blinded by a mêlée of excited hens.

'Hi!' I called out. 'I say there!'

'What the devil are *you* doing?' a rough voice replied, and looking up I saw a tallish man in gaiters at the gateway, with a younger fellow (looking pretty scared, I thought) behind him.

'I might well ask you that,' I answered, getting to my feet in no very good temper. 'Is it your custom in this part of the country to shoot at visitors without warning?'

'There was a fox,' the younger fellow said sheepishly. 'In the hedge yonder. They get after the fowls, see?'

'Fox indeed!' I said, and without another word I strolled across and retrieved my umbrella, which, to my great mortification, I found to be shot through and through.

'You tailored un good and proper, Fred,' the tall man said with a laugh.

'Somebody will have to pay for this,' I remarked sternly. 'It is a most outrageous thing.'

'You was trespassing,' the man said. 'Creeping about the hedges. Who are you, and what do you want?'

'See them birds pecking at his turn-ups?' the young hobbledehoy put in. 'He's mostly chock-a-block with grain, or something of that, if you ask me.'

'So that's the game, eh?' the other man said.

'My name is Wentworth,' I told them, sick and tired of this meaningless folly. 'I have an urgent appointment with Colonel Ripley. Be so good as to direct me to the Manor House at once, please.'

My manner must have made it clear to them with whom they had to deal. But even so I was obliged to show these two fools my letter from the Colonel before they would permit me to pass. 'Best call in at the farmhouse and see Mrs Jellaby,' the elder said finally, perhaps in a belated attempt to make amends. 'She'll maybe run you up in the Land Rover.'

One would have thought that I had had enough trouble and delay already, and might now hope that the final stages of this tiresome journey would be comparatively plain sailing. But I have often found that if ill-luck dogs one at the start of a day it is difficult to shake it off completely before evening. One thing leads to another, as they say. Had I not been opening and shutting my umbrella as I descended the steepish track into the farmyard (in order, of course, to see whether the ribbing had been as irretrievably ruined as the fabric), I dare say I should not have been attacked, or at least menaced, by what at first glance I took to be a bull. With the quick instinct of a countryman I made a sideways leap on to a kind of trestle or stand for milk churns that chanced to be at the side of the track, not knowing of course that it was in fact a wheeled trailer – still less that my weight would raise the shaft, or towing-bar, from the ground and set the contraption in motion.

Heigh-ho! It's a weary world at times. Tennyson and his precious arch keep recurring to my mind as I sit by my gas fire and ponder on the tricks that fate can play. To the Greeks, of course, it was overweening pride that led to man's misfortunes, but nobody, I imagine, will accuse me of that. Still, one must not make too much of what was, after all, no more than a gentle spill. Had the trailer overturned in the steeper part of the track there might have been a nasty accident, but luckily it kept going until we were fairly down on the more level midden, when the towing-bar met some obstruction and I was catapulted on to a heap of – well, straw, and so on. Two churns, one of which at least seemed to be practically empty, had already fallen off, and I suppose it was the considerable clatter they made as they rolled away into a cart-shed that brought a rather heavily-built woman in slippers to the farmhouse door.

'What's the idea of this, then?' she asked.

I was anxious, as may be imagined, to give a full explanation of my unceremonious arrival, but whether it was the partial wetting I had had while running for shelter earlier on, or whether it was some effect of grain dust, akin to hay-fever, I was seized by a most uncharacteristic fit of sneezing, and for some little time remained sitting where I had fallen unable to say a coherent word.

'Such a racket!' the woman said. 'I'd have thought it was the Last Trump, if there'd been lightning with it.'

'Atishoo!' I said.

'I don't know, I'm sure,' she said, coming nearer. 'Look at your trousers!'

'Atchoo – atchoo – *a-tisho!*'

'Have you come far,' the woman asked curiously, 'just to give that carry-on?'

I continued to sneeze for some time, while the woman made no offer to help, contenting herself with a series of wondering exclamations and the absurd observation that it never rains but it pours. But at last the paroxysm began to abate and I was able to speak, though not at first freely.

'My name,' I said, struggling to my feet, 'is Woo – dear me – my name is *Atcher* . . .'

'Is that your umbrella and all?' the woman asked.

I think it was the state to which my faithful old brolly (a present from my colleagues, as a matter of fact, on my fiftieth birthday) – it was the sad condition of my old friend, when I recovered it from beneath the trailer, that made it clear to me that I must abandon any idea of calling upon Colonel Ripley, for that day at least. I decided to cut my losses, and take myself off with the least possible delay. I had had about enough, to tell the truth.

'Perhaps you will be good enough to give a message to Colonel Ripley?' I began briskly. 'As you can see –'

'How you ever come to be on that trailer,' the woman said. '"Wagon Train" isn't in it.'

'Never mind that now,' I said. 'Please tell your master that Mr Wentworth was unable to keep his appointment today, owing to a combination of – a –, a –, – Confound it!'

'You'd better let it come,' she said.

'– owing to circumstances over which I had no control. Just tell him that, please. And that I – that Mr Woo – Woo – oh, devil take it, Woorasher! Say I shall be writing,' I shouted angrily. 'And good day to you.'

'I'll tell him you called,' the woman cried after me, and burst into peal after peal of totally unnecessary laughter. There has been a sorry decline in manners, I fear, even in the countryside.

So there it is. I suppose I should write some sort of explanation to Colonel Ripley, asking for another appointment, but I don't seem to have the heart for it, just now.

A Comfortable Billet

It has all turned out very well in the end, as is often the case if one keeps one's head and lets things take their course. Some men, I dare say, would have given up after the unlucky experience I had on my first attempt to keep an appointment at the Manor House, Stenshall, Wiltshire, but schoolmasters have to learn to take the rough with the smooth. A very courteous letter from Colonel Ripley, regretting any inconvenience to which I had been put and suggesting a date for a second visit, decided me to try again – and here, in short, I am, snugly housed in a bedroom twice the size of my little crib at Fenport, and nothing in the way of draughts to speak of considering the age of the house.

Needless to say, my second journey down here went off without a hitch. Mrs Ripley drove me from the station, pointing out this and that as we went along. She is a charming lady, and we very soon found out that she was at school with a Miss Soulby whose nephew was at Burgrove during my time there, though we could neither of us remember the boy's name. A strange coincidence, which helped, I think. She is the kind of person with whom one at once feels at ease.

So, in his rather more boisterous way, is the Colonel. He has broken a leg, poor fellow, and is obliged to spend the day on a couch in his study, but looks very fit and healthy none the less.

I naturally attempted an apology for my failure to get farther than Manor Farm on the Wednesday, but he brushed it aside with great good humour.

'Never laughed so much in my life, when I heard,' he said, slapping his good knee. 'Hens flying all over the place, shouting and shots in the five-acre, and then down the hill you come like a bat out of hell, balancing on a two-wheeled trailer by God, if Mrs Jellaby is to be believed, with oats pouring out of your ears and milk scattering this way and that like a Goddess of Plenty in her chariot – if I could have been *there* dammit! – and then *blam!* head-first into a heap of dung and a hundred and fifty-six sneezes to round it off. What an entry! "Mary," I said to my wife – didn't I, Mary? – "we must see this joker, if it's the last thing we do," I said, "and get the rea] inside story." Well, you know how it is. Mrs Jellaby seems to think you did it on purpose.'

'I dare say Mr Wentworth didn't find the experience very funny at the time, dear,' Mrs Ripley said in her gentle way, while I was considering how best to take this very exaggerated account of an admittedly absurd contretemps.

'It was certainly an unconventional way to arrive,' I said at last with a smile. 'But these things happen.' And I gave them a short account of what really occurred, to which I must say they both listened with a great deal of appreciation. But then I have always had the knack of telling a story, even if it is, up to a point, against myself.

'Lucky you didn't hurt yourself, my boy,' Colonel Ripley said, when he had had his laugh out. ' You must bill me for those trousers – and the umbrella, of course. We might have it mounted and hung up in the hall with the stags' heads and other trophies. What is it, by the way – an eight-pointer?'

I could not help joining in the laughter at this ridiculous notion, and capped the Colonel's fancy by suggesting that perhaps a plaster cast could be made of my ruined trousers.

This sally had even Colonel Ripley beaten, and after a little silence he said that perhaps he ought to come to the point and explain what kind of help he was looking for during his enforced idleness. It would be mainly answering the telephone, I gathered, and seeing to this and that, as his wife had to be out in the car, on farming and village matters, a great deal of the time. I replied that I should be happy to make myself useful in any way that was within my powers, and after some further conversation about detail, the thing was settled. They gave me a satisfying lunch of steak-and-kidney-pudding, which I always enjoy, and I left, promising to return with my baggage on the Monday. 'Let us know if you think of taking the short cut across the fields,' the Colonel called after me, but not being able to think of a suitable reply, I contented myself with a wave of the hand. The joke, in any case, is beginning to wear a bit thin.

So here, as I say, I am, safely back again at the Manor, with two very full and interesting days' work behind me. The Colonel has wide interests and keeps me busy on errands of one kind and another. Some of them might be thought a bit, well, *infra dig* for a man in my position, but I am no believer in making a fuss, especially as Colonel Ripley always remembers to preface an unusual request with the words 'Be a good chap,' which keeps things on a proper footing. Or so it seems to me.

It is a varied life. This morning, for instance, there was an order for linseed cake to be hastened, a note to take to old Mrs Coombes at the cottage, and a parcel to be got off. Then I had to ring Rogers at the Bull and tell him it was off ('Never mind what,' the Colonel said, in answer to my natural inquiry. 'He'll know'), give a message to the Rector about the Boys' Club and ask Mrs Jellaby how Phoebe was doing. 'Oh, and while you're in the village,' he went on, while I scribbled my instructions down on the pad I have bought myself (quite the secretary, eh?), 'you might look Mathers up and ask him about the insurance on Felicity. I want to be sure she is adequately covered. And get me a P.O. for 18s. 6d., will you? There's money in the drawer there. My wife has to be over at Sturminster all day in the car, I'm afraid, but there's an old bike of mine in the garage. It's got a basket, so bring a few leeks from the farm – only don't bring 'em to me, as you did with that fish yesterday, take 'em straight in to cook. There's a good chap,' he added, as my eyebrows rose.

There was also a message about a sack of potatoes that Mrs Ripley would deliver tomorrow, but I did not quite get the name of the person to whom I was to give it. Colonel Ripley gets rather impatient if one asks him to repeat things – it is his leg, I dare say, that makes him a little brusque at times – and in any case there was no need to bother him. The message was certainly to one of the people I had to call upon on other matters, so I had only to keep asking as I went along. That is what we used to call 'using your initiative' in the Army, where one soon learns to find things out for oneself. Colonel Ripley rather reminds me of my old C.O. in a way; he has a trick of taking it for granted that one knows what he is talking about, which of course is not always the case at first when the subject is an unfamiliar one such as Army Council Instructions or, as in this instance, livestock and so on. Had he made it clear to me that Felicity was a mare, which I could hardly be expected to know, my interview with Mrs Mathers (to take a case in point) would certainly have gone off a great deal more smoothly.

When a rather slatternly woman came to the door of No. 7, Cadnam Row, in answer to my knock, and told me that Mr Mathers was out, I naturally assumed that I was speaking to Mrs Mathers, as indeed I was. 'Was it something important?' she asked me, and I thought there could be no harm in passing on the gist of my message, particularly as I was under the impression that it concerned her.

'It was only that Colonel Ripley wanted to know,' I began '– the question is whether you are adequately covered.'

The woman flushed up and immediately glanced down at her attire, and following the direction of her eyes I realized that, in that sense, she certainly was *not*. To get over what might have been a momentary awkwardness, I decided to change the subject and, averting my eyes, asked her casually whether she was expecting a sack of potatoes.

'It's no business of yours what I'm expecting, nor when,' she cried furiously, and to my utter astonishment slammed the door in my face, giving me no chance at all to explain, as I was anxious to do in my own good time, that my opening question referred simply to insurance. These country people take a deal of understanding. It was a relief to move on to the Rectory, where I had a pleasant chat with the incumbent, a man of my own kidney with whom there was no need to fear misunderstandings or embarrassments. He told me much about Stenshall and its good people, which will be a help to me as I go to and fro, and in return I explained that I was acting as companion-secretary to Colonel Ripley, while he was laid up. 'Though I'm more of a glorified errand boy than a secretary, it seems,' I added, smiling to show that I did not really take my employment amiss.

'Well, it's something to be glorified,' he responded. 'I only live in hopes of attaining that status!' And on that friendly note we parted.

My final call was upon Mrs Jellaby, down at the Manor Farm, an encounter which, to tell the truth, I put off as long as I could.

'Well I never!' she cried, throwing up her hands in mock astonishment. 'If it isn't Mr Woo-Woo-Woorasher!'

I had half expected something of the kind, after my previous meeting with the lady, and deliberately ignored the impertinence. 'My name is Wentworth,' I said quietly. 'Am I right in thinking I am addressing Mrs Jellaby?'

'I can't forget it – ever,' she said. 'I was in the front, not to tell a lie, when the clatter starts up, and I said to myself "It's the atomic!" I said, "Or if it's not that," I said

'Mrs Jellaby! I have been asked by my employer, Colonel Ripley –'

'And sneeze!' the woman went on, wiping her eyes. 'Sitting there in the muck, kind of baffled, and not a word out of you but A-ratchoo till I thought to myself –'

'Mrs Jellaby!'

'But there! Come in do, Mr Wentworth,' Mrs Jellaby said, pulling herself together at last. 'I'm forgetting my manners. A glass of cider won't do either of us any harm.'

I was somewhat loth, as may be imagined, to accept hospitality after what had passed, but I had my mission to fulfil and took three glasses before I could get to the point. I must say that Mrs Jellaby proved, on better acquaintance, to be a very amiable woman and quite devoted to Mrs Ripley, as who is not? She is inclined to be a little voluble, perhaps, and utterly unable, like all these farming folk, to appreciate that what is clear to her may not be equally clear to an outsider.

'Phoebe?' she said at last, in answer to my repeated inquiries. 'We're not happy about her at all, not really. She's still not letting it down, tell the Colonel.'

'Dear me!' I said. 'Yes. I see. Not letting what down exactly, Mrs Jellaby?'

'Why, her milk,' Mrs Jellaby said, staring at me as though I were out of my mind. 'Whatever else, Mr Wentworth?'

'Of course, of course,' I said. 'I hadn't realized – that is to say, what precisely do you think is the cause of this – of the failure? Just in case the Colonel wants to know, you know.'

'They say I'm an out-of-date old silly, Mr Wentworth,' Mrs Jellaby replied, leaning forward very earnestly with her hands on her knees, 'but it's *my* opinion she's got a cold in her bag.'

'I see,' I said, wondering what some of my old colleagues would have thought of this extraordinary conversation. 'Yes. No doubt. One can only hope, in that case, that, unlike another lady I could name, she is adequately covered!'

I had her there! It was *her* turn not to understand what / was talking about.

A Cricket Dinner

'Where's that ass, Wentworth?'

I could hardly believe my ears when I heard my employer refer to me in this offensive way, and I dare say he could tell by my manner, as with a quiet 'I am here, Colonel Ripley,' I stepped into the study and stood waiting with my pad at the ready, that I was not accustomed to such treatment.

'Oh, there you are!' he said lightly. 'You mustn't mind me, Wentworth. Mary will tell you I call all my friends asses.'

If he expected to mollify me by the insinuation that I was now more of a friend than an employee he only partially succeeded.

'That may be. They are in a position to return the compliment,' I said warmly. 'I, unfortunately, am not.'

'You manage to make your point very nicely, all the same,' he replied with a good-natured grin, and feeling that honours were even I let the matter drop, particularly as Mrs Ripley gave me a conspiratorial wink, as much as to say 'You put him in his place very adroitly there, Mr Wentworth.'

'What's this I hear about your asking Mrs Mathers if she was expecting a sack of potatoes, you old rascal,' the Colonel went on, bursting into a roar of laughter. 'If *you*'d had eight children in nine years, and another on the way, I dare say –'

'Good heavens!' I cried, flushing to the roots of my hair. 'So that was why – I had no idea.'

'It's all over the village,' Mrs Ripley said. 'And nobody enjoyed the story more than old Mathers.'

The subject hardly seemed a suitable one for mixed company, I must say. However, it was not I that had brought it up.

'What an unlucky thing!' I said, really distressed. 'I shall certainly do my best to explain, and apologize, to Mrs Mathers.'

But to my surprise they both opposed this plan. 'You'd have to choose your words pretty carefully,' the Colonel pointed out; and with this, on reflection, I agreed.

It is difficult to avoid putting one's foot in it occasionally in these unfamiliar surroundings. Still, I am picking things up very quickly as I go along and in another week or so ought to be thoroughly *au fait* with farming affairs. By then, of course, my visit will be drawing to an end. The Colonel's leg is mending rapidly – 'I have you to thank for that, Wentworth,' he told me the other day. 'You are the best spur to recovery an invalid ever had' (a compliment that I greatly appreciated, as he is by no means over-ready with praise) – and as soon as he can get about my usefulness here will be over. I shall be sorry to go, in many ways, for they are friendly people hereabouts and I think I may claim to have made quite a hit with them. Everybody seems to know who one is, and so on, in a remarkably short space of time and to be anxious to stop for a chat. Only yesterday a total stranger hailed me to inquire whether my hay-fever was better, which I am sure was well meant, though I am not as it happens a sufferer.

As a matter of fact, it may well be convenient for me to leave in the fairly near future. Other things apart, I have had a belated reply to my advertisement in *The Times* from a Mr Bennett, of London, inquiring whether I might be free to accompany his two boys to Switzerland in about a fortnight's time. *In loco parentis*, I gather. One does not get away to the Continent very often on a pension like mine, and it seems too good an opportunity to miss, all being well. Still, all that is in the future. For the moment, there are things to be seen to in the outhouses! Upon my word, I sometimes wonder what the world is coming to. 'Be a good chap and nip out to the loft over the stables,' the Colonel began, so I knew there was something rather menial in the wind. How the boys at Burgrove would smile if they could see their old master checking over apples in a granary and putting the affected ones in a basket to take to Mrs Jellaby later – though what the good woman wants with rotten apples is more than I can say.

Tonight, apparently, I am to represent Colonel Ripley at the Annual Dinner of the Stenshall Cricket Club, of which he is President, so that one cannot complain of a lack of variety here. If it isn't one thing it's another, I said to Mrs Ripley in an unguarded moment, and she agreed. Her sympathy and understanding mean a great deal to me at times. She would have made an ideal headmaster's wife, had things turned out differently all round, with just the right manner towards parents, and it grieves me to see her carrying pig-meal about in an old pair of breeches. That is clumsily put, but the sentiment is sincere. 'Let *me*, Mrs Ripley,' I said to her on one occasion, and rather than hurt my feelings by refusing she handed over the bucket and thanked me very nicely. I think the unexpected attention touched her quite deeply, which explains why she went away without remembering to tell me where to take the pig-meal. There are no pigs here, as far as I know, so I had to get rid of it as best I could. The point is that her husband rather takes her for granted, in my opinion – not that it is any business of mine, of course.

Cricket holds the Empire together I told them, as one good fellow to another, and they liked it. Everybody sang afterwards, though I had had no warning, mind, not a word till it was too late and off I had to go. The Colonel knew. Ripley must have known what I had to do and I did it, no thanks to him. He never said. Anybody ought to, when people have to speak, but there it was.

'I am speaking on the President's behoof,' I said, which was true, despite the fact that I didn't know until I was told. What I mean is nobody told me before. 'Call upon Mr Wentworth to propose toast,' some fellow said, and there I was. I wouldn't have nine times out of ten. I said that, too, making no bones about it to clear the air. Nine times out of ten I wouldn't, I told them, but this is the tenth. 'Have some more cider,' they said, but I refused, unless that was later. I never drink more when I have to speak. Actually it was later I dare say because I didn't know I was going to before, as it turned out. I am scribbling on my bed while I remember, in case I forget what I was going to say. The Rector spoke, I remember that, and I spoke and another man spoke about a whip-round for bats ('In the belfry?' I said to my neighbour, but had to nudge him and repeat it, by which time the man – not this man, the other – was thanking the wives for helping with tea, so the point was lost). Where was I ? For two pins I would go to bed, but my pyjama top is missing. Play the game I told them and everybody cheered. 'Stand up the man who shot my umbrella!' I said, mixing *seria cum joco*, and they cheered again. Nobody stood up so I sat down, having no more to say at present. It is a rule I always follow when speaking, but they all shouted, 'Go on, go on!' and somebody said, 'You have forgotten the toast, sir.'

'What toast is that?' I asked, not having been told properly before, and the Rector said it was the Club, which was a great honour for one who had so newly come among them. Not the Rector, naturally. I mean it was a great honour for me, or so I told them. 'I am only the Colonel's legate, of course,' I said –

meaning to add 'his *broken* legate' for fun, but a bald-headed man got in first with the quip, and in the general laughter I could not think where in the world my pyjama coat can have got to. So I sat down again. Then the Rector proposed the toast of the Stenshall Cricket Club, and I made a short reply, which I forget. A Mr Binns told me I was the hit of the evening, but I ought to have been told before, in my opinion. Still, we all enjoyed myself and sang songs, which is the great thing.

It was under my shirt, of all places.

The days speed pleasantly by, with little of note to record. Colonel Ripley is up and hobbling about now, I am glad to say. Much of his old tetchiness has gone, too, which makes him easier to work with. Had he stepped into that puddle of pig-meal while he was still confined to his couch – not that he could have done so, naturally; I am merely drawing a comparison – he would have made a great deal more fuss about it, or I'm a Dutchman. Actually, it was a kind of mash for chickens, which I should have disposed of elsewhere had I known that my employer would be poking about in the shrubbery for some unexplained reason.

'Don't tell me *why* you put it there, Wentworth,' he said, before I had so much as hinted that I had anything to do with the matter. 'Some other time, perhaps, when I've an hour or two to spare. We must just be thankful you didn't send it to the Church Bazaar, along with that load of manure.' He was smiling as he spoke, or I should certainly have resented this unnecessary allusion to a perfectly understandable mistake. If the Colonel has a fault, as he undoubtedly has, it is a tendency to harp on trivialities that are over and done with. In any case I am no stenographer, and errors are bound to occur occasionally when instructions are rattled off faster than I can write. It was on the tip of my tongue to tell him that it was no part of a companion-secretary's duties to dispatch manure, to the Church Bazaar or anywhere else.

'Did I ever hear, by the way,' he went on, 'where my old hats went to?'

They went to Lady Wimbury at the Grange, as he very well knew, and as soon as she returned them with a short note I realized that there had been a muddle and took immediate steps to put things right. No harm whatever was done, as far as I can see, except for a little staining outside the village hall, which will wash off. It would have been different if the load had been delivered *in*side.

'It would, yes,' Colonel Ripley agreed, when I pointed this out. 'Now, be a good chap and give Mrs Ripley a hand in the stables, will you?'

Be a good chap, indeed! Be a good ostler and general factotum would be a likelier way of putting it.

Still, there are many compensations. I happened to be enjoying the sunshine on the terrace after luncheon when two or three shots rang out down by the farm. The men are rabbiting, I expect.

'Hullo!' I heard the Colonel remark, through his open study window. 'They're after Wentworth again!'

'Oh, I *hope* not,' Mrs Ripley replied, with her pleasant laugh. 'He's such a dear, really.'

'I wouldn't have missed him for worlds,' the Colonel said.

I moved away then, of course, being no eavesdropper. But I had heard enough. The sun went in after a minute or two, but I scarcely noticed it. There are things more warming than sunshine.

In Foreign Parts

The snow-capped peaks tower upwards and the lake shimmers in the bright sunlight. Had I the pen of a Ruskin I dare say I could describe the scene with more vividness, though of course there is less need for that kind of thing now that so many people travel abroad. After all, when one has seen a thing for oneself one does not much want to hear what somebody else thought about it – a point that modern writers often forget. One has one's snapshots, and so on, if memory proves treacherous, whereas in Ruskin's time I suppose even picture postcards were something of a rarity.

Still, it is certainly very pretty here on Lake Lucerne – or Vierwaldstattersee, as it is rather cumbrously called on the map (*Touristenkarte!*) provided free by the hotel – and I cannot help congratulating myself on my wisdom in advertising for temporary employment in *The Times*. There goes the steamer for Vitznau, a typically foreign contraption, absurdly broad in the beam for such calm waters, though it all adds to the fun in a way. We shall be aboard her in a day or two, I expect, when the boys have had a proper rest after the long journey. 'Take them about a bit and show them things,' Mr Bennett said to me at our last interview, and I feel sure that a trip round the lake would be in accordance with his wishes. Lucerne itself we must certainly see. Then there is the ascent of the Rigi, whence the views, so the hall porter tells me, are very fine. It will be best, I think, to draw up some kind of programme after lunch.

We have already been up one mountain, as a matter of fact, rather unexpectedly. Up to a point, that is. A man of my experience does not take a couple of boys all the way up a mountain by accident, I need hardly say. What happened was that on our way by train to Brunnen, where we are staying, from Zurich whither we had flown by the night plane, I distinctly heard a woman say that this was a Schnellzug (she was speaking in German, of course) and did not stop at Brunnen. As the train was then standing at Schwyz, which I luckily – or perhaps unluckily, as it turned out – knew to be the last stop before Brunnen, the last *station* that is to say, I bundled the boys out with our luggage in double-quick time, only to find from a most helpful official that the train which had now left did in fact stop at Brunnen and that the next one on would leave in an hour and a half. This was rather a facer, as we had not yet breakfasted, but the official suggested that we take a tram instead, from just outside the station, which we very soon did. I thought it best not to explain our little slip-up to the boys, as William was getting fretful and even his elder brother Geoffrey looked rather pale. Boys are happier, especially far from home, when they feel that everything is going smoothly, and a little harmless deception is often justified, in their own interests.

It was for this reason that, when the tram after a short run reached the centre of Schwyz and everyone alighted, I concealed my annoyance at having been misdirected and, with a cheerful 'Only one more change now,' simply got out and followed the rest into a waiting bus. Soon we were bowling along a delightful valley and though, had I been in my own country and a little less tired, I suppose I should have made more careful inquiries, I had no hint of trouble until a conductor came along flourishing his clippers and I demanded '*Drei nach Brunnen*', with a wave of my hand at the two boys. The conductor shook his head and said 'Stoos', of which I could make neither head nor tail, so I simply handed him some money which unfortunately turned out to be Italian. How it came to be in my right-hand trouser pocket I cannot think. It is my custom when travelling abroad to make a very careful distribution of money, documents and other valuables, to ensure that each is handy as and when required. Thus on the present occasion I had our passports safely in my inside breast pocket with the return air tickets pinned to the inner back cover. Travellers' cheques, as always, were in my left-hand hip pocket, which buttons, Swiss notes in my outside breast pocket secured with a safety-pin, and so on. Any English change I tie up in a handkerchief as soon as the frontier is crossed and keep for the time being in my left-hand trouser pocket. The small amount of Italian money I had brought with me in case we were able to make an excursion through the St Gotthard tunnel should have been in my top right-hand waistcoat pocket, and I got the shock of my life when in trying to explain all this to the conductor (with the aid of gestures, for he seemed a slow-witted sort of man) I put my fingers into the pocket in question and pulled out our return train tickets to Zurich! These should by rights have been in the front fob pocket of my trousers, but I soon gave up the attempt to make this point clear to the conductor, for after a brief glance at them he became so verbose and unintelligible that I had difficulty in keeping calm. Geoffrey began to ask what was the matter, and to gain time I blew my nose without proper forethought and instantly scattered a considerable quantity of small change about the bus.

Everybody was most helpful in hunting about under the seats, etc., for the coins, which to my astonishment proved to be Swiss.

'Then where is my English money?' I cried involuntarily as soon as I saw that the money they were handing to me was in francs. This started a fresh search among a number of the passengers who seemed to understand English, until I asked them not to trouble. 'It will turn up, no doubt,' I said. 'I must have put it somewhere else.'

'Perhaps it's up the other nostril,' I heard young William say in a whisper to his brother, but I was unable to reprimand him at once for the impertinence as an English lady just in front of me was asking whether she could be of any help.

'I thought I heard you asking for Brunnen,' she said. 'This bus goes to Schattli, for the cable railway up to Stoos, you know.'

'I see,' I said. 'Yes. And from there to Brunnen?'

'Well, you have to come back again to Schwyz, of course, and then there's a tram. But you might as well come right up to Stoos while you are about it. It's lovely up there.'

So that is what we did. The boys seemed to think it was rather a long way round to Brunnen, but they enjoyed the cable railway and became much more lively when we found breakfast being served in a fine hotel up on the top. The lady, a Mrs Fitch who is staying at Brunnen apparently, was very kind, drawing our attention to Lake Lucerne away below to our left and pointing out a number of the surrounding heights. She is a most friendly person, in the prime of life, and seemed to be as relieved as I was when I finally found my English small change while taking my tobacco pouch from my right-hand hip-pocket. 'You shower money from *every* quarter, Mr Wentworth,' she said gaily.

All that, however, is by the way. Here we are safely in our hotel, with the boys resting in their room on my instructions and I myself lazily watching the steamer grow smaller in the distance. It was a fortunate chance that Mr Bennett noticed my advertisement, and fortunate for me too that he had found himself unable, at the last moment, to accompany his two sons on their holiday abroad and so was obliged to look about for a trustworthy companion and squire for the lads. We very soon came to terms, and I must say that he has been most generous in his provision for the jaunt. It is quite a new experience for me to have another man's money in my pockets to spend (which reminds me that I really must get my small change and so on resorted and properly disposed or goodness knows what I shall be finding in my waistcoat next. Piastres, eh? Or a return ticket to Baker Street!). But I dare say I shall quickly get used to it.

Of course, it is a responsibility. But then, as I told Mr Bennett when he asked me whether I was used to taking charge – this was at our first meeting, to be fair, before he had had a chance to sum me up – I have taken parties of up to a dozen boys abroad in my time and not a broken leg between them. He looked a little dubious at this recommendation for a moment, but his brow cleared when I explained that I was referring to winter sports holidays. 'I see,' he said. 'Yes. I was not thinking of any physical danger so much. Geoffrey is a steady, sensible boy as a rule, but William – the younger boy, you know – is apt to run a little wild at times. I suppose you –'

'Oh, that!' I said, laughing. 'You must set your mind at rest on *that* score.' It was really too funny to think that I might be alarmed at the prospect of keeping two youngsters of eleven and thirteen in order. 'Why I have had fifteen of them at me at once before now,' I began, recollecting an occasion at Burgrove when the electric light failed – but, realizing that he was probably too busy a man to want to listen to stories of an assistant master's early days, I left it at that. In any case, by the day after tomorrow at latest he should have my postcard announcing our arrival after an uneventful journey, and his mind will be at rest.

News of our little diversion up the Stooshorn has somehow got round the hotel, and there has been some flattering comment on our energy in tackling a mountain before breakfast on our way out from England. 'Been up the Rigi yet, Mr Wentworth?' somebody called out as we made our way in to lunch, and a party of young Dutchmen began to sing 'There'll always be an England', *sotto voce*. I took it in good part, as one should on a holiday, but when I pulled out my handkerchief during the second course and some total strangers pretended to hunt about under their table I thought the joke had gone far off and spoke pretty sharply to young William for giggling. It is the first time he has heard the rough side of my tongue, and I think it surprised him.

It looks as though somebody has been making a story out of what was, after all, a very trivial mishap in the bus. I confess that I glanced momentarily at Mrs Fitch (who sits alone, I notice), but meeting her wide-eyed and friendly smile dismissed the thought as unworthy.

A Trip up the Rigi

I put on my pullover this morning, feeling in holiday mood, though one misses the extra pockets afforded by a waistcoat, and asked the boys at breakfast how they would like to make an expedition to the top of the Rigi, which seems to be the thing to do here.

Geoffrey asked me how high it was, and I told him I believed it was well over five thousand feet.

'Isn't that rather potty?' William said. He is at the age when they make rather a point of not being impressed and has to be taken down a peg or two now and then, in a friendly way.

'You must be careful not to trip over it when you aren't looking, William,' I said. 'That is, if you are thinking of making this trivial excursion on foot. Geoffrey and I are too old to make light of five thousand feet and will be going up by train.'

'In a ship first,' Geoffrey said. 'In a jolly old *dampfschiff*. Will Mrs Fitch be coming with us, Herr Wentworth?'

I suppose it is natural for boys to giggle at foreign words, though my own view is that that sort of thing should be kept for occasions when no foreigners are about whose feelings may be hurt. I contented myself, for the moment, with a slight frown to show that I was not greatly amused (often the best way with over-excited boys) and merely asked why he imagined that Mrs Fitch would be joining us.

'Well, she seems to. She came up the Urmiberg, I mean – and that other place, when we went up by mistake.'

'The Urmiberg catches the worm,' William put in, and both boys laughed so immoderately at this senseless pun, if such it could be called, that I was forced to check them. 'People are looking round,' I said, and led the way out of the breakfast room (or *Speisesaal* in the local lingo), humming a little tune as I sometimes do when a little put out. Not that William meant to be impertinent, I think. Boys are apt to say the first thing that comes into their heads, whether it means anything or not.

It was pure chance, really, that Mrs Fitch happened to be at the hotel entrance yesterday afternoon as we were setting out to walk to the cable-car that runs up to the Urmiberg from the outskirts of Brunnen. Naturally she inquired where we were going, and one thing led to another as it so often does. She is very well acquainted with the district and has already promised to show us another little railway that starts quite near the hotel and goes up to Morschach and Axenstein, whatever they may be. If she had said Rosencrantz and Guildenstern it would have been all the same to me. In any case, I see no harm in it at my age. More was lost on Morschach's Fields, I fancy! But I must not let the sunshine and heady air of Switzerland betray me into young master William's habit of punning, however apt.

How astonishing, in passing, is the industry of the Swiss! It seems that they cannot see a mountain peak without at once putting a railway of some kind up it. Often, too, they build a hotel at the top of it – which is natural enough, I suppose, when one considers the number of visitors who make the ascent. Unless, indeed, the hotel was there first, which would certainly make the construction of the railway or funicular more understandable. To take people up to the hotel, I mean.

But, in that case, it is hard to see how the materials and so on were taken up to the hotel in the first place. It is like the chicken and the egg, as I said to the hall-porter when discussing the problem with him after breakfast; but he does not, I think, understand English as well as he would have one believe. He simply directed me to the *Speisesaal*, from which of course I had only just come. I thanked him gravely, and to spare his feelings pretended to glance over the letter rack until he went away. One says 'thank you' for useless information a good many times a day, I find, when abroad. It is all part of the game.

We had a splendid day on the Rigi. The journey up the lake to Vitznau is most inspiring, with ever-changing views of the surrounding mountains and other points of interest, which I did my best to point out to my two charges. But they were generally on the other side of the boat, as boys so often are. So I chatted pleasantly with Mrs Fitch, who to my surprise turned out to be on board, and let my conscience go hang. After all, although it is my duty to see that this holiday is educational for them, in the fullest sense, it is no bad thing to let them broaden their minds on their own once in a way. I am always there, if they need me.

I had not realized, to tell the truth, that Mrs Fitch was a widow, until she told me that she had no one to care for now. I was wondering what to reply when the ship put in at Gersau, to pick up more passengers, and no sooner was this distraction over than Geoffrey came across to ask a question about my map, which he had very sensibly borrowed. So the opportunity was lost. However, as I had not been able to phrase any entirely suitable words of commiseration or consolation, no very great harm was done, I dare say. It is only that one likes not to be thought boorish or indifferent.

'The point is,' Geoffrey said, flattening the map on the seat beside me, 'this is supposed to be a lake, isn't it?'

'It is indeed,' I replied. 'With a maximum depth, just about where we are now I believe, of over 700 feet. So don't go leaning too far out over the rail, you two, or I shall be after you.'

'Then why is it called a "*See*"?' he demanded, stabbing his finger on the map. 'Look here, where it says Vierwaldstattersee. *See* means "sea", surely. It's swanking.'

'My dear boy,' I chuckled, after a quick look round to make sure that his remark had not been overheard, 'you really must not try to tell these good people what they should call their lakes. Switzerland has no sea coast, as you know, so why should they not use the word "sea" for such waters as they have? And very fine waters they are, too, are they not?' I concluded, raising my voice.

'They made a good job of it, anyway,' William joined in, jabbing at the map as his brother had done. 'Look, it's the Vierwaldstattersee where we are now, and the Urner See where it goes round the corner at Brunnen . . .'

'*And* the Kussnachter See in that sort of creek thing,' Geoffrey said, 'and the Luzerner See or something there, and the Alpnacher See down here. Golly, it's a swizzle.'

'Five seas in one lake,' William counted. 'It's a good thing they haven't got a real sea or they'd soon run out of names.'

'And what is your definition of a real sea, young feller-me-lad?' I asked, to test him.

'Something that's salt anyway, and doesn't keep going round corners,' he declared, and we all burst out laughing.

'That would include Lot's wife, wouldn't it?' Mrs Fitch asked innocently.

'Or a bloater,' I added, to keep the fun going.

With one thing and another it was a light-hearted voyage, and so hot into the bargain that I had to go below before we reached Vitznau to remove my pullover. We had another good laugh as soon as we were ashore, when Geoffrey pointed out an

English poster at the railway station saying 'Come to Skegness!' or some such place, and William wanted to know why on earth the Swiss should want to go *there*. 'Perhaps they like to look at a sea that doesn't go round corners,' I said with a twinkle, though I thought it right to add, when the laughter had died down, that young people must learn to admire the beauty of other countries without belittling their own. Mrs Fitch, for some reason, chose this moment to tell me that I was an unbelievable pet, a remark which I hope the boys did not overhear. They sometimes misunderstand grown-up teasing.

At Rigi Kaltbad, about two-thirds of the way to the actual summit, we left the train for luncheon, on Mrs Fitch's advice, and fed very comfortably on a terrace off some excellent veal steaks. We elders drank a bottle of lavaux (a local wine I am told) – an expense that I think Mr Bennett, my employer, would have approved, though I shall of course offer to reimburse him for Mrs Fitch's share of it. The strong air made me feel very fit. Looking out over the vista of mountains and down to the gleaming lake below it seemed a far cry from Dora's Café in Fenport, where I sometimes have a cup of tea, and even from Burgrove School, happy though I have been there at times. A line or two of poetry, summing it all up, hovered at the back of my mind, but when I turned to share the thought with the others I found I could think of nothing but an oddly-worded notice about the boys' dirty laundry that was once put up on the School Board by a temporary matron. The mind works in curious ways at times.

And so to the summit, the Rigi Kulm itself. 'My goodness!' I could not help exclaiming, when the magnificent view burst upon us, 'What an amazing sight.' And everybody within earshot agreed with me. The boys simply stood and drank it all in, while I quietly spelt out the names of such peaks as I could identify from the map and reminded them how with the aid of trigonometry the height of each one of them could be accurately determined. Boys are always interested, in my not inconsiderable experience, to be told how the lessons they are learning at school are practically applied in the larger world outside, and I was bring-them up gradually from sea-level to a series of gradually established trig, points when the inevitable tiresome bore, whom none of us knew from Adam, intervened with the announcement that on a really clear day it was possible to see the Flugelhorn, or it may have been the Gruntstock.

'Indeed!' I said coldly. 'That must be delightful. But we can see all the peaks we need as it is, thank you.'

I fear I was a little abrupt, but really! He took himself off almost at once, but the spell was broken, and we too made for the train, to complete the round trip back to Brunnen, via Goldau and Schwyz. 'Well, well, well, well, well!' I said, smiling round at the others, as the train jogged and jolted its way down between rocky walls. 'We shall all be glad of a cup of tea when we get in.'

They were silent, however. It seemed odd that they should have nothing to say after what had been, when all is said and done, a remarkable experience.

A Glimpse of Italy

It is no surprise, naturally, to find so many foreigners in Switzerland, but one had expected them to be Swiss. Or, to look at it from the Swiss point of view (as I make a point of trying to do when abroad – when in Switzerland, that is to say, and of course, *mutatis mutandis*, elsewhere) one had expected the foreigners to be *English*, which is what we are really, as I keep reminding the two boys, when we leave our own country. I mean it is we who are the foreigners, not the other way round, whereas in fact most of the others here seem to be Germans and Dutch and Danes and so on. Quite a cosmopolitan gathering, and very different from the old days when the English were the great travellers. Still, it all adds to the fun and has certainly opened the eyes of my two young charges. 'As you see, the English are not the only people in the world,' I sometimes say to them, to drive the lesson home.

To give them their due, both Geoffrey and William have been keen to make the very most of their continental holiday. 'When are we going to spend that Italian money?' they kept on asking me before we had been here a week, and in the end I agreed that we would make a little expedition through the St Gotthard tunnel into Italy. Just for the day, of course. I had had this plan in mind all along, as a matter of fact, and had the forethought to bring a few thousand lire with me, a fact of which the boys happened to be aware. But *having* money does not necessarily mean that one must *spend* it, as I felt it my duty to make clear to them. Besides, it does them no harm to be kept on tenterhooks for a while. Discipline is not always so easy to maintain while on holiday as at school; one can hardly set them impositions and so on, in cases of disobedience. But I had only to say quietly, 'If you do that again I shall not take you to Italy, William,' and that was the end of it for half an hour or more. 'When you've a treat in store for the little devils,' as a wise old colleague once said to me, 'keep it up your sleeve as long as ever you can.'

I decided that Lugano would make a pleasant jaunt, and we caught the ten-forty-three from Brunnen, arriving at our destination after a thoroughly Alpine run at twenty-three minutes past one. This gave us a good two hours in the place before taking the half-past three train which gets back to Brunnen at six-twenty-five, in good time for dinner. The boys enjoyed the great tunnel, which I confess was a new experience for me too, and eagerly pointed out to each other the Italian names on the farther side. The sun seemed to shine with added warmth as we ran down the southern slopes of the mighty range and I think I dozed off for a while. At any rate the time passed quickly, and we were all in high spirits and ready for lunch as we left Lugano station and strolled to the lakeside in search of a suitable place to eat.

'Look, *ristorante*!' cried Geoffrey. 'How super!'

Then William spotted the word '*impermeable*' in a shop window full of raincoats, Geoffrey capped it with '*pantaloni*', and not to be outdone I silently pointed a finger at the absurd legend '*pizzicàgnolo*' which happened to catch my eye. 'What's it mean?' William demanded, and I was still considering my reply when Geoffrey suddenly stopped dead in his tracks and said excitedly, 'I say, Mr Wentworth. They never stamped us in.'

'Stamped us in?' I retorted. 'Whatever do you mean, boy?'

'Our passports,' he said. 'You know. Nobody bonked them,' and he banged a fist down into his open palm to show what he meant.

'Nor they did,' William put in. 'Gosh, what a swizzle!'

'Well,' I said, laughing, 'what of it? I dare say as we are only here for a couple of hours –'

'How should *they* know?' Geoffrey said. 'Besides, I *want* it bonked.'

'There's no need to get in a state about it,' I said. 'And please talk sensibly.'

'So do I,' William shouted. 'I want it to say Italy. There's not much point otherwise.'

I was inclined to tell them both not to be silly little fools. If they thought I had brought them all this way simply to have their passports stamped they were no better than babies. But on reflection it did seem a little odd that we had somehow missed the frontier formalities, and of course there was the possibility that when we attempted to re-enter Switzerland there would be a fuss, since we should have nothing to show that we had ever left it or where we had been in the meantime.

'Very well,' I told them. 'If it is as important as all that we must see what we can do,' and I began to retrace my steps, with the idea of inquiring at the railway station. However, we soon spotted a policeman, or *carabiniere* as they say, and to save time I put our little difficulty to him. We had come, I explained, from Switzerland, just for the day, but through some oversight our passports had not been stamped on entry. Would he kindly advise me how to regularize the position?

'*Passaporto?*' he said.

'Yes,' I replied, speaking very slowly and clearly. 'Not stamped. *Sapristi?*'

It was soon clear that he understood little if any English, and having myself no great command of Italian I took my passport from my inside breast pocket and opened it at the page bearing the Swiss entry stamp at Zurich. 'See?' I said.

'*Si*,' he replied.

'Now then,' I went on, squaring my shoulders, 'here' – and I laid a finger on the entry stamp – 'is Switzerland. *Schweiz*. Yes?'

'*Si*,' he said. '*Svizzera*.'

'As you will,' I replied. 'But here,' jabbing my finger on the empty space below, 'no stamp! *Marka? Indorsimento?*'

He nodded his head several times, but I have spent too many years as a schoolmaster not to recognize a look of utter incomprehension when I see one and I therefore brought my clenched fist sharply down on the open passport. '*Bonk!*' I said, shaking my head to indicate that it had not been done.

'Try *bonka*,' one of the boys advised. 'They always put an "a" on it.'

The officer smiled vaguely, looking up and down the road as though for help, and I began again at the beginning. '*Schweiz*,' I said pointing. '*Svizzera*. Yes?'

'*Svizzera*,' he said. '*Si*,'

'Good!' I said. '*Buona!* But Italy – *Italia* – no!'

'*Italia*,' he agreed, 'no!'

'Well then!' I cried in exasperation. But it was useless. The policeman, with a polite '*Scusa!*' took my passport from me, flipped over the pages, studied my entry visa for America for a moment, and handed it back with a bow. 'Come!' he said, finally.

We followed him, willy-nilly, for a few hundred yards until we came to what I surmised, rightly, to be a police station, where to my great satisfaction I very soon found myself talking to an officer who spoke, on the whole, very good English.

'But Mr Wentworth,' he said, raising his eyes from my passport, when I had briefly explained the situation, 'Lugano is in Switzerland. There is no need for your passports to be stamped until you cross the frontier.'

'Then we are not in Italy?' I cried, unable to believe my ears. 'God bless my soul! But everybody speaks – the signs – I always thought the St Gotthard . . .'

'We also speak German in Switzerland, as you know, and we are not, I am happy – we are not, that is to say, in Germany. We speak French, in the Valais, and we are not in France. Yes?'

'You also speak excellent English,' I said warmly, 'and you are not –'

'Exactly,' he said. 'Though that is rather different.'

I cannot think, looking back, how I came to make so stupid a mistake. But there it was, and I could only be thankful that the two boys had remained out of earshot during our conversation. All the same, I was not out of the wood by a long chalk.

'My two young charges will be sadly disappointed,' I told the official, after apologizing for so unwarrantably wasting his time. 'They had set their hearts, for some reason, on having Italy stamped in their passports.'

'So?' he said, smiling. 'It is always the same, when we are young. But it is only a few kilometres to the frontier, on the Menaggio road. Would it not be possible –?'

I glanced at my watch. 'Unfortunately,' I said, 'we have to catch a train back to Switzerland – to Brunnen, that is to say – at half-past three. And we have not yet lunched. These formalities, in my experience, take time.'

'I see,' he said. 'Yes. One moment, monsieur.' He left me to confer briefly with one of his colleagues, and returned with an expression of the greatest delight. 'All is arranged,' he said. 'Please to follow me.'

Well, to cut a long story short, the good-natured fellow took the three of us out to the frontier post and back. In a police car! 'Gosh, we're doing eighty!' young William shouted, and so we were, though it was only in kilometres of course. 'You are really extremely kind,' I said to our friend, but he merely smiled and said, 'Leave everything to me,' which I was only too glad to do. We were stamped out of Switzerland and into Italy in the twinkling of an eye, and as we stepped from the farther post on to Italian soil I could not resist the temptation to raise my hat in the air and cry gaily, 'Well, boys, what do you think of that?'

'Whacko!' they both said. 'But it seems funny –'

'Time to be going, if you want any lunch,' I told them, and in a minute and a half, all told, with many friendly grins and shouts of 'Stay a bit longer next time', we were stamped out of Italy again, '*Bonk, bonk, bonk!*' as William aptly put it, describing our whirlwind progress. 'And one *bonk* to come,' his brother added, as we approached the Swiss frontier post once again.

Here, however, there was an unexpected hitch. Two customs men sternly demanded whether we had anything to declare – any silks, leather goods, watches, cameras, etc., and before I had time to say a word one of them made a dive for my waistcoat pocket and dragged out my gold half-hunter. 'Italian, no?' he said. 'You bought her in Milan just now?'

'That watch belonged to my great-grandfather, young man,' I replied with some heat, 'and what is more –' But I suddenly noticed that everyone was laughing, including our policeman friend, and there was nothing for it but to join in. What a people, eh?

So there it was. The four of us lunched together, at my expense (or rather at my employer, Mr Bennett's, strictly speaking, though I am sure he would have been the first to authorize the extra expenditure, had he been with us), and a very jolly meal we made of it. 'Any time you are passing through Fenport, Hampshire,' I said to our good friend when the time came for us to part, 'don't fail to look me up'; and he promised that he would, or wouldn't, rather. 'I really mean that,' I added. And so I did.

'The funny thing *is*,' William said sleepily as our train climbed up towards Airolo, 'how when we were in Italy already we had to sort of go out of Switzerland to get into it again. Into Italy, I mean. And then we went out of Italy and into Switzerland, so as far as I can see we ought to be in Switzerland *now*.'

'So we are, William,' I said.

'But in that case –'

'Oh, stow it, Bill!' his brother said, and though I do not in the ordinary way allow rudeness to pass unchecked, even between close relations, I left it at that.

Scotched Rumour

It is good to be back in one's own country again, with solid, sensible words like 'Family Butcher' in the shop windows instead of that absurd *pizzicàgnolo* (which means 'pork butcher', I am told) and policemen in proper helmets. One misses the mountains, of course, and the air, which has been likened, I forget by whom, to champagne. But I feel very fit after the holiday, and the two boys benefited I am sure both mentally and physically. 'They tell me they never had a dull moment in your company,' Mr Bennett very kindly wrote when sending a final cheque which will be more than useful. 'I only wish I could have been with you to share the fun.' Speaking for myself I am thankful that he was not, since I should not have been there had he been able to accompany us. But no doubt he meant it kindly.

Having no further engagements at present I am settling down into the old routine at Fenport, and have no intention yet awhile of renewing my advertisement for temporary employment in *The Times*. One needs a little rest at my age, greatly though I have enjoyed my varied experiences. My three weeks in the heart of Wiltshire, followed by this fortnight abroad, have amply repaid the trifling original outlay, both in money and health. I could eat a horse, as they say.

The second-hand furniture nuisance has died down, I am thankful to say. None too soon, for really it was very difficult to know how to get rid of all the people who brought odds and ends for me to see, and my little cottage is seriously overcrowded. I suppose it was weak of me to buy an egg-collector's cabinet, for I am no oölogist, but at eight and sixpence it was something of a bargain and may well come in handy in the long run. I have a plan for removing the partitions and filing receipts and so on in the drawers. Or I may present it to Burgrove, if the Headmaster approves. We shall see.

Meanwhile the most ridiculous rumours are going about here. My own concerns are no business of anyone else's, and I could see no necessity to tell people here that I was seeking temporary employment, but my absences have been noticed, of course, and the gossips have been busy. 'I hear you were caught trying to smuggle second-hand watches out of Italy,' Mrs Wheeler astonished me by saying, in the full hearing of the tobacconist. 'Was it in aid of this antique business of yours in Wiltshire – or Northumberland, is it?' What a farrago of nonsense! It was all I could do to keep my temper.

'Really, Mrs Wheeler!' I exclaimed. 'I did not think that you, of all people, would listen to such silly talk. If Miss Stephens has been going about –'

'Oh, Mr Wentworth!' I was dismayed to hear Miss Stephens herself exclaim. 'How could you?' She must have followed me into the shop without my noticing, and I had no option but to apologize.

'I am sorry if I appear to have been talking about you behind your back, Miss Stephens,' I said stiffly. 'But upon my soul, the tittle-tattle that goes on in this place is enough –'

'There's so little else to do, you see,' Miss Stephens said. 'We can't all go dashing off to the Continent on mysterious errands. You shouldn't be such a dark horse, Mr Wentworth.'

'Dark horse, indeed!' I began, fingering my tie. 'It is true that I have been in Switzerland for a short spell –'

'Well, I'm glad you don't deny *that*,' Miss Stephens said. 'Or I should have to confound you by producing Myra Fitch's letter.'

'Mrs Fitch!' I cried, colouring despite myself. 'Well, well, well. Good gracious me!'

'She is an old friend of my mother's,' Miss Stephens said, 'and Mother sent on her letter about her holiday. She thought it might interest me, as it said quite a lot about a man from Fenport. It did.'

'Aha!' put in Mrs Wheeler, to my great annoyance. Mrs Fitch is a very charming lady, who was good enough to help me look after the boys on one or two of our expeditions in the mountains, and naturally we spent my last evening together on the hotel terrace, admiring the tranquil waters of the lake and talking over this and that, as one does. Had I been ten years younger, I dare say – who knows? – I might have suggested a stroll along the shore, but the night air grows a little chilly after dinner. But it is no business of anybody's that I can see.

'An ounce of Richmond Curly Cut, please, Mr Gooch,' I said loudly, to show the two ladies that I had no wish to continue the conversation, which seemed to me quite unsuitable for a tobacconists's shop. ' And a box of matches.' But it takes more than a box of matches, however briskly ordered, to persuade Miss Stephens to leave well alone.

'I can give you her address, if you like,' she told me, with an archness that I found most distasteful.

'I have it, thank you,' I replied, before I could stop myself, and raising my homburg with as much politeness as I could muster bade them both good morning. The whole affair is trifling, but it is difficult to get it out of one's mind. One cannot help wondering, naturally, what Mrs Fitch said in her letter. Nothing to my discredit, I'll wager.

So her name is Myra, eh?

To quieten the spate of talk I have thought it best to let it be known in the district that I have recently taken one or two temporary posts, to eke out. One would have preferred to keep this sort of thing to oneself, but something had to be done to scotch the rumour (started, I shall always believe, by Megrim) that I had become a kind of Queen's Messenger and was likely to leave for Ankara at any moment. Another theory I heard being discussed in the Post Office was that I had business interests abroad – I wish I had – and that 'this second-hand furniture racket', as one man had the impertinence to call it, was only a blind. It was getting beyond a joke. Anyway the truth is now out, and the result is that I am being bombarded with surreptitious gifts of fruit and vegetables and so forth. People are very kind, but really! I have a sufficiency of means, as long as I am reasonably careful, and much as I have always enjoyed *Cranford* (one of my favourites since I was quite a young man) I do not at all relish the role of a Miss Matty, however well meant the thought behind it. They will be setting my cottage up as a sweet-shop next! It is hard to know what to do, without giving offence, and in any case one cannot return a cauliflower left, without message of any kind, in the tool-shed. Even when I caught Mrs Wheeler red-handed putting a dozen eggs down by the back door she managed to make a tremendous favour of it. I mean a favour to her, of course. 'I'm simply snowed under with the wretched things, Mr Wentworth,' she told me. 'So if you could *bear* to help me out. . .' I tried to pay for them, although I already had four dozen in the larder, but she said it was more than her life was worth under the Egg Marketing Scheme or some such extravagant rubbish.

I don't know, I'm sure. It occurred to me to make repayment in kind to some of these good people, and I did manage to leave a small gate-legged table in the Wheeler's garage and a pair of brass-bound bellows at Miss Stephens's place, besides one or two other bits and pieces which relieved the congestion in my sitting-room. But it was difficult work after dark, and little good came of the plan in the end. All the pieces were recognized and returned, either direct to me or via the people who had sold them to me while the secondhand furniture rumour was at its height, and the general belief (which I had not the heart to deny) was that the whole thing had been the work of a practical joker, who had raided my cottage and distributed my belongings at random. 'And a joke in damn bad taste, too,' Mr Wheeler remarked to me with a good deal of heat. One way and another there was quite a fuss, with everybody doing everything they could to make it up to me. It is all rather embarrassing, and I often wish I had some sympathetic soul, like Myra Fitch, to talk things over with.

I thought it only civil, in passing, to send her a card – or brief note, rather; nobody who knows the postmistress here would send a postcard even on quite ordinary topics, as was mine I need hardly say – to inquire about her journey home among other things, and she replied with a very kind letter, which filled my head with what might have been dangerous thoughts in a less balanced man. What have I to offer a lady after all, even if the thing were feasible, except a small pittance and an outhouse full of vegetables and corner cupboards? Ah well! There it is. I wrote a long reply, not wishing to seem unfriendly, and as luck would have it ran into Miss Stephens on my way to the post. 'Caught you, Mr Wentworth!' she cried gaily, eyeing the envelope in my hand with a great deal of curiosity. 'There are going to be some sad hearts in Fenport when the announcement appears in *The Times*.'

I might have made a sharp reply, had I not caught a glimpse of what looked suspiciously like tomatoes in her bicycle basket. All this kindness, awkward though it is in a way, is quite disarming.

'The only announcement you are likely to see in *The Times*, Miss Stephens,' I said with a smile, 'is that Mr Wentworth is no longer available for temporary employment in positions of trust as he is setting up a greengrocer's shop in Fenport.'

It was rather naughty of me, I suppose, but her confusion was delightful to see. 'Caught *you*, I may say, Miss Stephens!' I added. 'It is very, very kind of you all, but it must really stop. I am quite all right, you know – quite all right,'

'Well,' she said, after a little silence, fiddling with her handlebars, 'we happen to be rather fond of you in these parts, you see. Still, I'll tell them.'

'My dear young lady,' I began, but found it quite impossible to complete whatever it was I had been about to say.

'You aren't going away on any more of these jobs, I hope,' she said. 'Not for a bit, anyway.'

'No, no,' I assured her. 'No. I shall spend the next few months among all my good friends in Fenport.'

I was wrong, however. By the very next morning's post I received to my astonishment a letter from my old Headmaster, the Rev. Gregory Saunders, M.A., telling me that the school was in some difficulty owing to the sudden illness of Mr Thompson and asking whether I would consider tiding them over by returning to Burgrove for the last five weeks of term. 'You shall have your old IIIA mathematical set. Do come!' he wrote.

Well! It did not take me long to make up my mind, as may be imagined. Back to the fray on Monday, eh!

I felt quite boyish as I sat polishing up my mathematical instruments after tea. It will be like old times to feel a piece of chalk between one's fingers again.

Back to Burgrove

It is grand to be back here again, if only 'to tide us over' as
the Headmaster puts it. Well, I have tided them over a few
difficulties in my time, and I dare say I can do so again. Of
course it is not the same. One cannot expect to step back into
one's old seat at the top of the Common Room tree on the
strength of a few weeks' temporary work. Easy does it. 'Tact,
Wentworth old boy!' I said to myself as I shaved this morning.
'Tact and diplomacy!' And be sure I shall need both. I had to
hold myself in pretty tight directly after chapel when I was
hanging up my gown in the old familiar cupboard. They've put
in a new light-switch, I noticed: one of those pull-down things
on the end of a long string, which always seems to me a bit –
not suggestive exactly. Anyway I don't like them. But it wasn't
that. It was a young fairhaired fellow, new since my time and
takes French and History they tell me.

'I say,' he said, 'you must be Thompson's stand-in. That's
Mr Rawlinson's peg, if you don't mind my telling you. He's bit
touchy, you know.'

Well!

It was on the tip of my tongue to inform this young hopeful
that the peg in question happened to be mine, that I had used
it for twenty-seven years (ever since the Lent Term 1933, when

old Poole gave up, to be precise), and that if anybody was trespassing it was Rawlinson. But my sense of humour won the day and I thanked him instead, saying with an assumption of the utmost gravity, 'I am most grateful for the hint. You have saved me from an irreparable blunder.' I then took my tattered old gown off the peg and hung it up again, with mock humility, on the farthest peg of all, right in the corner where the rolled-up map of Europe before the Great War used to stand.

'That's mine,' the young fellow said.

Still, all the familiar smells are there, and I snuff them up like an old war-horse returning to the fray. The Headmaster, in particular, has been most kind and welcoming. 'It is like a breath of fresh air to have you back with us, A. J.,' he told me, to which I replied, jokingly (though to tell the truth I was very much moved), 'The School hasn't often been short of fresh air, surely, Headmaster?' – a reference to the central heating system, which was always going wrong in my time. But he missed the point, I think. He has grown rather fat in late middle age, and is no longer known to the boys as the Squid, so Rawlinson tells me. Apparently they call him the Atomic Pile, in their modern way, or 'Tommy' for short, though his real name is, of course, the Reverend Gregory Saunders, M.A.

I had quite a shock on entering Classroom 4 for my first period with my old mathematical Set IIIA. The lower part of the wall, a sensible dark green in the old days, has been painted primrose, of all colours, with a lighter shade above, on some cock-and-bull theory that boys work better in cheerful surroundings. Nonsense! Boys work best when they have got their heads down over their books, with a master in charge who knows how to keep a firm hand on the young rascals, not when they are staring at fancy plastic emulsions. I suppose it is all part and parcel of turning the place into an 'Inspected School', which happened as soon as my back was turned. We are to have a second visit from these gentlemen in a week or two, the Headmaster tells me, and much good may it do us, or them. It is difficult enough in all conscience to teach boys the Theorem of Pythagoras, without being distracted by some Government popinjay sitting in judgement on the teaching methods of a man old enough (though by no means inclined) to be his father.

However, what was in some ways an even greater shock awaited me with IIIA. One's first duty, naturally, is to list the boys' names. Not that they are not already listed in the mark-book by one's predecessor, but it makes a start and helps one to get acquainted and so on.

'Call out your names, please, one by one,' I told them, 'beginning from the left of the front row.'

'Do you want them in the form order, sir?' somebody asked.

'Naturally,' I said. 'That is why I said beginning from the left.'

'The top boy sits on the right, sir.'

I was thunderstruck. Boys at Burgrove sit at their desks in the order of the previous week's mark-lists, and in all my experience it has been the rule that the top boy sits on the left, the next boy on his right, and so on down the rows, ending with the bottom boy (who has to wipe the board and do other small chores) at the extreme right of the back row. Any other arrangement leads, in my opinion, to nothing but confusion.

'In my classroom,' I said, 'the top boy sits on the left. Now will you please get yourselves sorted out in the proper order as quickly as possible. And *quietly*! This is a classroom, not an elephant-house.'

It is extraordinary what an amount of noise a dozen boys can make with their feet, but eventually, after I had given a pretty sharp look to a biggish dark boy whom I caught tweaking another boy's ear as he passed, they all settled down again, and I began to write their names in my book as they called them out.

'Henderson,' I repeated, 'Blake, Wrigley . . . With a "W"?' I asked, looking up at the third boy in the row.

'Yes, sir. As in Wrekin.'

To my astonishment it was the boy next to him who answered, that is to say the third boy from the *right* (there being seven desks in all in the front row, as I ought perhaps to have made clear), and I immediately demanded an explanation. 'Has Wrigley lost his tongue?' I asked sharply. 'Or why do you take it upon yourself to speak for him?'

'I *am* Wrigley,' the boy said, looking genuinely bewildered.

'I see,' I said. 'Wrigley, did you not hear me say that you were to sit in your form order *beginning from the left*. Can you not count up to three?'

Wrigley simply stood there, looking helplessly about him, until the boy on the extreme right, who turned out to be Henderson, kindly put his oar in. 'I think I can explain it, sir,' he said. 'Wrigley thought you meant our left, not yours. We all did, sir. That's why I'm over here where I am now, instead of being where I was when I started, if you see what I mean, sir.'

It is most important that a master should be fair, as well as firm, and believing that there had been a genuine misunderstanding I said no more than 'Very well, Henderson. But understand this, all of you. When I say "left" in this classroom I mean *my* left and nobody else's. Is that clear?'

'What happens if you are speaking with your back to us?' somebody asked.

'Stand up the boy who said that,' I ordered, in the voice I use only when I mean to have no nonsense. A fair-haired boy with glasses, whose face seemed vaguely familiar, rose to his feet a good deal more slowly than he will learn to do when he knows me a little better. 'I only meant –' he began.

'Your name?' I said sternly.

'Mason, sir.'

'Mason!' I repeated. 'Indeed! Mason, eh? Well, well, well, well. Good gracious me! I see. How old are you, Mason?'

'Eleven and a half, sir.'

There was a fair-haired boy called Mason here in the old pre-war days, with whom I crossed swords on one or two occasions. Not a bad-hearted chap, but a little too inclined to overstep the mark. Indeed at times he was downright insolent, which I am scarcely the man to tolerate. It would be odd, though not of course impossible, if I were now to have the doubtful pleasure of trying to cram the elements of algebra into his son's head.

'May I sit down now, sir?'

Some of the other boys laughed at this, and I very soon spotted the reason.

'You appear to be sitting down already, Mason,' I said. 'So I am afraid I fail to see the point of your question.'

'Oh, so I am sir. I must have done it without noticing. What I mean is, may I have your permission to sit down, sir?'

If I had not been quite certain before, this sort of tomfoolery was enough to convince me of the boy's identity. I was anxious to have no unpleasantness at my very first period with IIIA, but the sooner this youngster was put in his place the better it would be for all of us.

'Mason,' I said slowly, 'I believe your father – All right, boy, sit down now – your father was at this school, I believe, in the late 1930's. Is that so?'

'Yes, he was, sir. He told me all about you.'

'Indeed!' I said. 'That must have been very interesting. And did he tell you, among all the other things, that I was not a good man –'

'Oh, no, sir.'

'Not a good man,' I continued, raising my voice, 'to try to be funny with? Do you happen to remember that, Mason?'

The boy had the impertinence to pretend to be racking his brains, until I brought him to his senses by rapping sharply on my desk with a pair of compasses.

'I expect he did, sir,' he said hurriedly. 'Sir, is it true, sir, that you once fell backwards into a kind of basket in the boot-room?'

'Be quiet, all of you,' I cried. 'We are wasting far too much time. Henderson, where had you got to with Mr Thompson before he became ill?' I had forgotten, until the boy I was addressing told me his name was Sibling, that the form was still back to front owing to this misunderstanding about left and right, and there was a further tiresome delay while they all got themselves back into their original positions.

And even then, as Henderson reminded me, I had still to take their names down before we could get started.

'Henderson, Blake, Wrigley with a "W",' I said briskly – 'those I have got. Next?'

There was no reply, and I had to repeat the order. But the silence continued.

'Come along, come along, wake up!'I said. 'You, there – what is your name?'

'Kingsley,' the fourth boy said, looking frightened, as well he might. 'But I'm not really next, sir.'

'Then why are you sitting there, boy?' I thundered, beginning to lose patience. Upon my soul, I began to wish I had my old IIIA lot back again, muddle-headed as many of them were.

'It's Potter who's next in the order really, sir,' Henderson explained. 'But he isn't here.'

'Why is Potter not here?'

'I don't know, sir. I think he had to go and see Matron.'

'See Matron and die,' somebody sang out. I suspected Mason, but in my profession one has to be on one's guard against prejudice. So I let it go, and went on with my list of names as though I had noticed nothing. Which was just as well, as it turned out; otherwise I might not have got to the last boy before the bell rang for the end of the period.

We must really get down to it tomorrow. Still, the hour was not entirely wasted. As every schoolmaster knows, it is of the first importance to get on terms with one's boys. Let them see what they are up against right from the start, and then – off with a bang!

Getting the Feel of It

Boys always want to know what is the *use* of what one is trying to teach them. Thus they are always more ready to find out how much wallpaper is needed to decorate a room than, let us say, to factorize an expression or follow a theorem involving tangents. I suppose it is natural enough, in a way, and I do my best to point out some of the applications of mathematical laws and principles to modern life. But one has to be on one's guard against being led astray into time-wasting digressions. One tip, that may be helpful to young men just starting out on the long furrow, is to watch your step when the boys begin to lean back and cross their legs. It means that the young rascals are counting on a good ten minutes' break from real work, and is one of the danger-signals that should never be disregarded. I make it a rule to get straight back to x and y as soon as I see it.

This morning's discussion with IIIA about the diameter of the sun (it originated, rather oddly, in a problem we were working on together about the price of eggs) went on perhaps a little longer than I generally permit, but that was because some valuable lessons seemed to me to emerge in the course of it. The importance of angular measurements, for one.

'Was there ever a time, sir,' Wrigley asked, while I was writing some solar dimensions on the board (and finding incidentally that chalk still breaks on the down stroke just as it always did), 'when people thought the sun was no bigger than the earth?'

I could not resist telling them of some of the weird misconceptions of the ancients, and in particular of a remark made by the Greek philosopher Epicurus which has always amused me, namely that 'the sun is about as large as it looks, or perhaps a little smaller'. This made them all laugh, with the exception of young Notting, rather a serious-minded boy for his age.

'Well, isn't it?' he asked.

He seemed genuinely puzzled, so I told the others to be quiet.

'What I mean, sir,' Notting went on, 'if it looked bigger than it does it *would* be bigger, wouldn't it, and if it were smaller than it is it would *look* smaller, so it must be as big as it looks.'

It is not always easy to follow the mental processes of young boys, so I took a turn about the classroom to clear my mind, confiscating a mint humbug from Mason as I passed his desk.

'And how big does it look to you, Notting?' I asked at length.

'Well, I suppose actually,' he replied, holding up a forefinger and thumb, 'about that.'

'Exactly,' I said with a smile. 'About an inch across. And now what about *my* diameter? No, no – that will do, Mason – how big do *I* look? Measure the width of my shoulders in the same way, with your fingers. So!

'You had better do the same, all of you,' I went on, when Notting had satisfied himself that his mathematics master was three times as big as the sun. 'Extend your arms and make a note of my breadth as it appears from where you are sitting. Then we shall soon learn something about the relationship between distance and angular measurement.'

They at once complied, and I was about to warn them to take the measurement with thumb and forefinger, as Notting had done, and not between their first and second fingers, when the Headmaster looked in to ask some question about cod-liver oil and malt, which I should have thought could have waited till break. He raised his eyebrows on entering, frowned at the boys, and was somewhat curt in his manner when putting his trivial query. There are times when I have scant patience with these headmasterly moods. He should take more exercise, in my opinion.

I was proceeding, after his departure, to demonstrate on the board from figures supplied by the boys how the width of an object can be determined by the apparent width of its extremities at known ocular distances (though I put it more simply, of course) when I became aware of a faint hum, or buzz, tending to increase momentarily in volume.

'Who is humming under his breath?' I demanded, swinging round on my heel.

Nobody owned up, though the humming continued, and I confess I derived a certain grim enjoyment from the thought that this old dodge should be tried out on an experienced hand like myself. It is almost impossible to pin down the actual culprit – culprits, rather, since more and more boys join in if the thing is allowed to continue. The lips do not move, and the utter immobility of the boys often has an unnerving effect on young masters. I remember a young Cantab, Griffiths was it? Or Fenner? Some name of the sort – anyway he did not last long – who reached the point of imagining that the noise was inside his own head and actually took pills for it, until Gilbert took him aside in the Common Room. Still, that was all a long time ago.

'There will be a serious row,' I told them, 'unless this humming stops immediately.'

Looking at their bland, expressionless faces (though one or two of the older ones attempted a puzzled wrinkling of the brows that might have fooled me thirty years ago) I was irresistibly reminded of my old IIIA of pre-war days, and very near burst out laughing despite my annoyance. 'You have had your last warning,' I said.

'I think it must be an aeroplane, sir,' Mason said.

'No doubt,' I said. 'Could it be the same one your father heard in similar circumstances, perhaps?'

A boy called Barstow in the back row had the impertinence to call out,' It's coming nearer!' and somebody said he thought it was a Gipsy Major, or some such nonsense, by the sound of the engine.

'Very well!' I cried, bringing my fist down with a crash on my desk. 'Since it is impossible to work while this noise continues, we shall all have to make up the time lost after school hours this afternoon.' And to ram the warning home I ostentatiously took the time by my watch.

Instantly, as I had expected, the humming ceased.

'He's crashed,' Mason said – a remark that he may yet regret, when I have made up my mind what to do about it. For the moment my main concern was to waste no more time and get down to some solid work straight away.

'Open your books at Exercise 37,' I told them brusquely, 'and get to work on the first six examples. No, Hopgood, you may not. Wait till the bell goes, boy. If there are any difficulties – Well, Blake, what is it?'

'It's the aeroplane, sir.'

For the life of me I could not help following his pointing finger and there, sure enough, was a plane of some kind. However, it was only a second or two before I recovered my poise.

'I see,' I said. 'How curious that we should hear its noise so long before it arrived, Blake.'

'It's the speed of it, sir. The waves get left behind or something.'

'I know all about supersonic effects, thank you,' I said sternly. 'The only difficulty about your ingenious theory is that the sound arrives afterwards, not before.'

'This must be a subsonic plane, sir,' somebody put in. 'So it's the other way round.'

'Unless it's flying backwards,' Mason said.

The bell went just then, and saved one or two of them from what might well have been the surprise of their young lives. For the time being, at any rate. Later on, we shall see. I mentioned this aeroplane business to Gilbert during break, thinking it would amuse him to hear how badly this new generation of boys had mistaken their man. He tells me that the plane flies over regularly at about 10.30 every weekday morning and his guess is that the boys take advantage of it, but misjudged the timing on this occasion. Well, well, well! It begins to look as if this man Thompson, whose place I am temporarily filling, were an indifferent disciplinarian, who has allowed the Set to get a little out of hand. It may be necessary to tighten things up for a day or two, until they learn where they stand.

Hopgood, it appears, is no relation of the Hopgood II whom I so unfortunately stunned with a *Hall and Knight* back in 1939. This is a relief, really, as I have no wish to have that absurdly exaggerated affair dragged up again, should the boy's parents happen to come down. Not that I have anything to be ashamed of or regret, except that the wrong boy was hit. It is just that the thing is over and done with, as I told Rawlinson pretty straight after luncheon, when the talk happened to turn on the sons of Old Boys. I was a bit put out already, to tell the truth, by an extraordinary remark of the Headmaster's a little earlier. 'I assume,' he said, taking my arm, as we walked down Long Corridor together, 'I *hope*, Wentworth, that it was the Victory salute your boys were giving you this morning. But whatever it was, don't you feel it would be better –'

'We were discussing the diameter of the sun, Headmaster,' I interrupted with some impatience, 'and I really feel that in such matters my judgement is at least as good as that of a classical man. My methods of work, which are founded as you will agree on a great many years' experience –'

'Classical I admit to being,' he said. 'But even a man whose training has been mainly in the groves of Academe may perhaps claim some knowledge of the significance of gestures and the advisability or otherwise of their concerted use by boys during working hours. Or not, Wentworth?'

'Gestures!' I cried, with some heat. 'I know nothing of gestures. I happen to be more interested in the angle subtended by the sun –'

He did not, however, allow me to complete what I had to say.

'God bless my soul, A. J.,' he cried, slapping me with sudden friendliness on the shoulder.' You have the power to astonish me, even now!'

Having no idea what he was driving at, I took myself off in something of a huff. Still, I think he means to be kind, and it is pleasant, taking one thing with another, to be back in the dear old place.

A Brush with the Inspectors

It has been a tiring week, on the whole. No doubt Her Majesty's Inspectors have their work to do, and for all I know there may be schools that benefit from it, but I could wish that they would do it elsewhere than in my classroom when I am busy. It is a vexatious thing that this visitation should coincide with my brief stint of temporary work here at Burgrove, particularly as the Headmaster has allowed it to throw him completely out of his stride. Fuss, fuss, fuss! Really, for the best part of a week before these gentlemen were due one hardly knew whether one was on one's head or one's heels. Quite apart from all the sweeping and cleaning, which I dare say was overdue, the school piano has been moved five times to my knowledge and is now back in the music room where it started. I should have thought, as I remarked to Gilbert, that if these people want to see the school at work they should see us when we are fresh and at our best, not tired out with re-marking all the boys' hairbrushes in the downstairs washroom. Then, the usual issue of clean blotting-paper, which has always been on a Monday for as long as I can remember, was held back for three days, despite my protests ('They'll hardly give us many marks for industry,' I pointed out, 'if everybody's blotting-paper is as white as driven snow on a *Thursday*,' but nobody would listen to reason), and to cap it all

I got wind of a plan to reorganize my own time-table without consulting me.

Of course I went straight to the Headmaster.

'I am told,' I began without preamble, 'that I am not to be permitted to take IIIA in geometry on Thursday morning, but that for some reason I fail to understand they will go to Gilbert for General Knowledge at that time. In all my long experience –'

'Certain rearrangements have had to be made as you know, Wentworth,' the Headmaster said, 'in view of the Inspection on Thursday and Friday –'

'That certain rearrangements *have* been made I know well enough,' I put in bitterly. 'Whether they have *had* to be made is another matter.'

'You will kindly leave that to my judgement,' he replied, in a tone I was inclined to resent. 'I have to consider what is best for the School. It is essential that the Inspectors' report should be a favourable one, particularly with regard to standards of discipline and methods of teaching.'

'Well?' I said, as he paused. 'I have yet to learn what all this has to do with the reorganization, without my knowledge, of my normal Thursday time-table, or,' I added (for I am a great believer in striking while the iron is hot), 'with the appearance in my classroom this morning of a new check duster, which will not rub out, in place of the old yellow one which will.'

The Headmaster sank into his chair with a weary sigh, confirming my growing suspicion that he is no longer really up to it and should begin to think of handing over to a younger man. 'I had supposed,' he said at length, 'that you would be glad to be spared the worry of having to teach with an Inspector in the room. However, if you insist on making an – if you prefer to be inspected, so be it. So be it, Wentworth. Only, for goodness' sake, don't come bothering me about dusters at a time like this, there's a good chap.'

I had half a mind to tell him that if he had not bothered about dusters in the first place I should have had no occasion to mention the matter. But he looked overstrained and in no condition to argue sensibly. Besides I had gained my main point. So with a quiet 'Worry? About an Inspector in my room? I'll soon settle *his* hash, if I'm not greatly mistaken,' I went off to see about the school roller, which had to be moved behind the cricket pavilion for some inexplicable reason. Upon my word, there could scarcely be more of a to-do if we were getting ready for Parents' Day.

As a matter of fact a very unsightly bare patch was left in the grass beside the gymnasium when we had moved the roller, and I was at my wits' end to find something to cover it up until Gilbert suggested moving the roller back there again. 'It ought to be just about the right size,' he said with a grin, and though I suppose I ought strictly to have referred the point to the Headmaster it seemed simpler to act on my own initiative. After all, if the decision was delayed for any length of time there would soon be a bare patch behind the pavilion and we should then be between the devil and the deep sea, as the saying goes. So I called up my helpers again, and as luck would have it we ran into the Headmaster as we were trundling the thing back to the gym again. 'Good, good!' he called out. 'That's the spirit!' Which only goes to show. As far as I can see there is no coherent plan at all. So long as everything is being moved from one place to another and back again we are doing all we can to prepare for the inspection. It reminds me of the Army, in a way.

'Required to prove,' I announced, throwing a piece of chalk light-heartedly into the air and catching it again before it struck the floor, 'that the exterior angle of a triangle is equal to the sum of the two interior opposite angles. Yes, Potter?'

I had decided to run over the familiar theorem again, not for the benefit of Mr Edwards of Her Majesty's Inspectorate who was sitting at the back but because it is only by constant reiteration that one can implant the basic principles of geometry in young minds.

'I don't see how a triangle can have an exterior angle, sir.' Potter said. 'I thought the whole point of a triangle was that all its angles were inside.'

In the ordinary way I should probably have told young Potter to wait until I had set the whole theorem out on the board before raising difficulties, but this morning one had to make allowances. I had, as a matter of fact, in a short talk the day before, warned them all not to be afraid to ask questions just because a stranger was present, as it would have given an entirely wrong impression of a lively and not unintelligent Set if they had all sat silent and mumchance throughout. So that Potter was only trying, according to his lights, to carry out my instructions.

'It is true, Potter,' I explained, tossing the chalk up again, 'that a triangle is, by definition, a plane figure bounded by three straight lines – never mind that now, thank you, Henderson; I have another piece – and in that sense may be said to have interior angles only. But if, in the triangle ABC which I am drawing here, I produce BC to any point D, will you not agree that I have made an angle ACD which may fairly be called an exterior angle?'

'I suppose so, yes, sir,' Potter agreed, in that grudging way so typical of boys.

'But it isn't a triangle any longer,' Mason objected. 'It's more of a corner-flag lying on its side.'

I joined in the laughter – which I dare say surprised the Inspector, who is probably more used to the sort of class where boys have to be kept continually on a tight rein for fear of indiscipline. It would do him no harm to see that a friendly relaxed atmosphere is by no means impossible, if the master knows what he is about.

'You remind me of your father, Mason,' I observed jokingly. 'He, if I remember rightly, disapproved of a triangle with squares on each of its three sides, on the grounds that it looked more like three squares joined together with a space in the middle.' I could not resist stealing another glance at Mr Edwards as I said this, to see how he was taking the intimation (not altogether unplanned, I fear) that he was sitting in judgement on a man old enough to be teaching a second generation. But he was looking out of the window with an abstracted air, and it was all I could do not to ask him to be good enough to pay attention.

'In any case, Mason,' I went on, 'if you dislike my producing BC to any point D –'

'To *any* point D, sir?'

'Yes.'

'I see, sir.'

'– to any point D, you will be even more distressed by my next step, which is to draw GE parallel to BA – so.'

One gets a little out of practice, in retirement, at the difficult art of drawing on the board, and I snatched up the duster with the idea of re-drawing CE more nearly parallel to BA, only to find that I had hold of this wretched new check affair, full of dressing, which did nothing more than make an ugly smear in the middle of my diagram.

'Botheration!' I cried, as I think anybody might in the circumstances. 'Where is my old yellow rag?' And I threw the offending article aside on to my desk where, by a cruel mischance, it knocked over a vase of daffodils, which had no business to be there in any case. If I had noticed them before I should certainly have had them removed out of harm's way, but one cannot have eyes in the back of one's head when concentrating on teaching.

'Who put those things there?' I demanded, and as nobody answered I put the onus on the top boy. 'Henderson,' I asked firmly, 'do you know anything about this?'

'No, sir,' he said. 'At least – I think there are flowers on all the masters' desks this morning, sir.'

'What in the name of goodness,' I began – but happening to catch Mr Edwards's eye I decided, out of loyalty to the Headmaster, not to complete the sentence. 'I see,' I said. 'Yes, yes. Of course. I was forgetting it was Thursday!' Which was rather a neat way, I flatter myself, of conveying to the Inspector the impression that it was quite the normal thing to have flowers in the classrooms on at least one day in the week.

'Are we going to have them every Thursday in future?' the little blockhead Wrigley asked, and though several of the boys immediately said 'Sh!' the harm was done. I must say, I blame the Headmaster chiefly, for trying to curry favour with these absurd fal-lals, and I only wish he had been present to see two of my boys attempting to mop up the mess with his infernal duster. 'It seems to be waterproof,' one of them rightly said, and in the end I had to send him out for a towel, which turned out to be not much better. 'I'm afraid it's brand new too, sir,' the boy complained. 'I don't know what's come over the place' – upon which I confess I could not for the life of me help exchanging a covert grin with the Inspector. What nonsense it all is, to be sure.

This little mishap unavoidably took up a good deal of time before I was ready to re-draw my parallel line CE, and I could not quite see what Mr Edwards was getting at when he took me aside later in the day and expressed his disappointment that in the fifty minutes he was with me I had not managed to reach the proof of the theorem I set out to explain.

'The proof?' I said. 'There is no great difficulty about that. Since the lines CE, AB are by construction parallel, the angle ACE is equal to the interior opposite angle BAG, and the exterior angle DCE – I have an envelope somewhere, I think –'

'Yes, yes,' he said. 'I am aware of the proof, Mr Wentworth. It was not myself I was thinking of but the boys.'

'Oh, the boys,' I said, laughing. 'Don't worry your head about the boys. They know it backwards by this time. Or if they don't they never will.'

'I see,' he said. 'In that case I find it a little difficult – perhaps you wouldn't mind telling me what you feel to be the basic principles of teaching young boys.'

'Basic principles!' I repeated, raising my eyebrows. 'There are no rules of thumb or short cuts to success in my profession, Mr Edwards. There is only one essential ingredient that I know of – a lifetime's experience.'

'Put *that* in your report,' I was tempted to add, 'and smoke it.' But I didn't, naturally. Nothing is to be gained by antagonizing that kind of person.

End of Term

'Well, well, I suppose there can be no harm in it now.'

The boys had been pressing me to tell them something of the old days at the School – the 1930s seem like the beginning of the world to them, of course – and as it was the last lesson of term, when one traditionally relaxes a little, I had half a mind to indulge them. Mason was particularly anxious to know why the Headmaster used to be nicknamed 'the Squid' – a fact that he learned from his father, I don't doubt – but I was certainly not prepared to go into that. I was never very clear about it myself, to tell the truth, unless it had something to do with the octopus's habit of concealing itself behind a cloud of ink. Certainly, Mr Saunders used to put up an unconscionable number of notices on the board in his earlier days, mainly about bootlaces and the wearing of caps on school walks and similar trivia. But then again, I don't know. These things just happen, as often as not.

'I was only wondering,' Mason said, when he saw my frown. 'Of course, if it was anything you wouldn't like to –'

'That is not the point, Mason,' I said. 'One does not discuss the Headmaster in the classroom, as you very well know.'

'Like sex and religion,' somebody whispered: an impertinence that I should have come down on like a ton of bricks had it been a normal working period. I am not easily shocked, but some of these youngsters nowadays seem to me altogether too old for their years.

'The School was a very different place when I first came here as a young man,' I went on smoothly, hitching up my gown, 'and by no means so bright and comfortable as it is now.' And I told them, after cautioning two boys for whistling, how there was no linoleum in the upstairs passages and only six basins in the washroom for nearly seventy boys.

'There are only ten now,' Wrigley said.

'And a hundred and eight of us,' Blake added. 'If you count Hopgood, that is.'

'You shut up, Admiral,' Hopgood said, momentarily forgetting my rule that all remarks must be addressed to me. 'I wash a jolly sight more often than you do.'

'Where the carpentry shed now stands,' I continued, quelling these interruptions with a look, 'there was in the old days nothing but – What is the matter now, Mason?'

The boy was bowed over his desk in a half-crouching position and appeared to be engaged in some kind of a struggle.

'My tie's caught. In the hinge, sir.'

'Open your desk then and free it, boy,' I ordered. 'Though how in the world –'

'I can't, sir. The lid catches me under the chin.'

'Excuse me, sir, but your gown's hitched up.'

'I am aware of that, thank you, Potter,' I said coldly. 'Now, Mason, I give you exactly ten seconds to get that tie free and sit down properly, or there'll be serious trouble. Ten seconds, mind!'

'Can I do the count-down, sir?'

I took up a piece of chalk and flipped it into the air. 'We can easily spend the rest of the period on parallel lines, if you prefer it,' I warned them, taking a significant pace towards the blackboard.

'Oh, but, sir! Then we shall never meet,' some fool called out, and I should certainly have returned to normal teaching then and there if the rest of the Set had not promptly told the offender to be quiet and begged me to continue with my reminiscences.

'Please go on, sir. Sir, tell us about the old days.'

'Sir, there was a scimitar in the Museum . . .'

'Tell us about when you threw the hot-water jugs at Matron.'

'Sir, is it true you shot the Bishop of Tewkesbury single-handed?'

'Oh, stow it, Coutts. I want to hear about the boot-basket.'

'Sir! Sir! My father said you were looking for your umbrella on a fine night . . .'

'That will do,' I said sharply. 'Quiet everyone, please. Wrigley, do I strike you as the kind of man who would throw water-jugs at anyone, Matron or anyone else?'

That silenced them, as I knew it would, and I took advantage of the pause to unhitch my gown and say a few straight words to the Set. 'Every school,' I told them quietly, 'has a lot of silly, exaggerated legends about the past, and it appears, I am sorry to say, that Burgrove is no exception. I make allowances for a certain amount of over-excitement on the last morning of work, but I have no intention of permitting my classroom to be turned into a bear-garden. If you cannot sit quietly and sensibly – Do you want to leave the room, Notting?'

'No, thank you, sir.'

'Then why is your hand up?'

'I wanted to know if I could ask a question, sir.'

'Very well,' I said patiently.

'Thank you, sir.'

The boy said no more. Indeed he casually picked up a pencil and began to doodle. So after staring at him for a minute or two in dead silence I rapped smartly on my desk. 'Get on with it, Notting,' I told him. 'We haven't got all day, you know.'

'Who? Me, sir?' the boy exclaimed, looking up with an innocent air which did not deceive me for an instant. 'I'm sorry, sir. I haven't got a question ready yet, actually. I only wanted to know if I could ask one in case I happened to think of one later, sir.'

'He wants to sort of bank one, I think, sir,' Mason was kind enough to put in, in his interfering way.

I thought I knew every dodge for wasting time that boys can get up to, but this was a new one even to me. However, I was more than equal to the occasion. If he wanted to cross swords with me he would soon find that two could play at that game.

'I see,' I said, without raising my voice. 'Very well, Notting. I am delighted to know that you believe in making provision for the future. Perhaps, that being so, you will have the goodness to write out "I must not try to be funny in class" fifty times – just in case I happen to want to set you an imposition later, you know.'

The other boys roared with laughter, and I must say it was all I could do to keep a straight face myself at Notting's comical expression of discomfiture. Somebody called out 'Sucks to you, Notty!', and though the expression is one I generally jump on I let it go this time. It seemed to me to sum up the situation rather neatly.

Needless to say there was no further trouble, and all the boys listened attentively while I told them the true story of the time the whole School went to the Tidworth Tattoo – or *should* have gone to Tidworth, rather. The bell rang while I was explaining how my own party were somehow misdirected to Aldershot, so the dénouement will have to wait for another time. If there is another time, of course.

'Well, Notting,' I said, as I rather sadly collected my books for the last time. 'What do you think now of your question-banking scheme?'

'Not much, sir,' he had the honesty to admit, and feeling that he had learnt his lesson I let him off the fifty lines. After all, end of term doesn't come every day.

Bur grove! Bur grove! Through the ages
Boy and master sing your praise!
Turn, yea, turn the crowded pages,
Ne'er forget those happy days!

How true it all is, I reflected as we sang the old song together at End-of-Term Supper, though of course 'through the ages' is stretching it a bit, as Gilbert says, for a school founded in 1907. Still, there is such a thing as poetic licence, is there not? I for one shall not forget 'those happy days', and tomorrow, when I am alone with my own thoughts again in my little cottage at Fenport, these last few snatched weeks at Burgrove will seem like a dream, I dare say.

All the same, nothing is to be gained by indulging in nostalgic self-pity, as though one were an old man with all one's life behind one, especially as the Headmaster has just told me that it looks very much as if Thompson would be away for several months yet, so that, should I care to consider it . . .! I shall certainly turn the matter over in my mind and let him know in a day or two. Temporary assistance, as the Headmaster pointed out (he does not scruple, bless his heart, to use every possible means of persuasion to get me back), is devilish hard to come by these days. 'I wouldn't dream of asking you, if I knew where else to turn,' he said to me; and though the sentiment was clumsily put I knew very well that he was thinking only of the sacrifice of well-earned leisure that my acceptance would entail.

In any case I am not likely to be dull during the holidays. As a Vice-President of the Football Club there will be this and that to do, no doubt. Then, of course, if I decide to come back here, I really must brush up my algebra a bit, which has become surprisingly rusty with disuse. I was very near floored the other day by a problem about the average speed of two cyclists, of all things! And I should not be very surprised if Miss Stephens is after me again about her precious Dramatic Society. Life is very far from being over! In fact it seems to be richer and fuller, in many ways, since my so-called 'retirement'. What with Switzerland and the Ripleys and one thing and another, not to mention Inspection Week here (though I am quite prepared to forget *those* particular 'happy days', School Song or no School Song!), one has been in quite a whirl.

Talking of Miss Stephens, incidentally, I gather from Mrs Fitch (a lady whom I met in Brunnen, as I may have mentioned), that she (Myra Fitch, that is) hopes to come down to Fenport shortly with her old friend Mrs Stephens (who is Miss Stephens's mother, naturally) to stay with her. To stay with Miss Stephens, I mean. Anyway, she (Mrs Fitch) tells me in her last letter that she is very excited at the thought of seeing me once more.

It seems a strange coincidence that we should meet again. One can hardly suppose – And yet, I don't know. We shall see. There are times when I scarcely know what to think. In certain eventualities it might even be that my decision whether or not to return to Burgrove would ultimately depend on – well, on circumstances. At my age one is not a callow youth. Nor is one a doddering old man, with one foot in the grave, if it comes to that.

So there it is. Or may be, rather. Back to Burgrove for one more term, or – an engagement of an altogether less temporary kind? I don't know, I'm sure. Even supposing. But it is nice to feel that one may yet be of some use, in one way or another.

Preview

It is to be A.J. Wentworth's final appearance on the scholastic scene. Once more he dons his cap and gown – or, to be more precise, Rawlinson's cap and gown – and returns to Burgrove for just one more time.

His final term includes a brief but broadening visit to the United States, in addition to the usual intellectual cut and thrust of the classroom. Whether he's causing a stir on Fifth Avenue, or merely 'trying to knock a bit of sense into a bunch of thick-headed boys,' A.J. Wentworth fumbles, blusters and generally carries on.

The Wentworth Papers, Book 3

Also Available

THE PAPERS OF
*A. J. Wentworth,
BA*

'One of the funniest books ever.'
SUNDAY EXPRESS

H. F. ELLIS

There is chalk in his fingernails and paper darts fill the air as A.J. Wentworth, mathematics master at Burgrove Preparatory School, unwittingly opens the doors that lead not to knowledge but to chaos and confusion.

In his collected papers he sets out the truth about the fishing incident in the boot room, the real story about the theft of the headmaster's potted plant, and even the answer to the sensitive question of whether or not Mr Wentworth was trying to have carnal knowledge of matron on that one, memorable occasion.

The Wentworth Papers, Book 1

OUT NOW!

About The Wentworth Papers

A classic comic study in blinkered English manners, the Wentworth Papers was first introduced to readers in the pages of *Punch* magazine. It was later dramatized for both BBC Radio and ITV drama.

The full series –

The Papers of A.J. Wentworth, B.A.

The Retirement of A.J. Wentworth

The Swan Song of A.J. Wentworth

About the Author

Humphry Francis Ellis was born in 1907 in Lincolnshire, and educated at Tonbridge and Magdalen College, Oxford. Following a year as assistant master at Marlborough school he began to write for *Punch* magazine.

In 1949 Ellis became *Punch*'s Literary and Deputy Editor, a post which he held until 1953. It was during this period that he developed the character of A.J. Wentworth, inspired by his experience as a schoolmaster.

Punch continued to publish Ellis's work, though from 1954 he found a more lucrative market in *The New Yorker*, where the Wentworth stories proved very popular.

Note from the Publisher

Milton Keynes UK
Ingram Content Group UK Ltd.
UKHW031835161024
449700UK00002B/6